The Author

Dr. Raj Kumar, Scientist-D and Head Radiation Biotechnology group, Division of Radioprotective Drug Development and Research (DRDDR), Institute of Nuclear Medicine and Allied Sciences (INMAS), Defence Research and Development Organization (DRDO), Delhi did his post-graduate in Microbiology from Gurukul Kangri University Haridwar in 1994. Dr. Raj Kumar completed his Ph.D. in Biotechnology from Indian Institute of Technology Roorkee (IIT-Roorkee) in 2001. In the same year he joined INMAS, DRDO, Ministry of Defence, Govt of India. He has worked on different aspect of radiation biology and radioprotection and explored plant *Podophyllum hexandrum* for its radioprotective properties. Later on he shifts his interest on radioresistant bacteria and their application in radioprotector development. Currently he is working on development of radioprotective drug for human application using radioresisnat bacteria. He has discovered a novel radioresistant bacteria *Bacillus* sp. INM-1 and isolated a novel efficacious radioprotective molecule from it. He has been awarded for various DRDO awards like "Scientist of the year award-2007", "Technology group award-2007" and "DRDO Commendation award-2015" for his significant contribution in area of radioprotective drug development. Till date he has published ~60 original research papers in International Scientific Journals of repute in the area of radioprotection. He has also contributed 7 book chapters in different books published by reputed International publishers. His effort is continue to cherish the long standing dream of DRDO to develop an efficacious, non toxic radioprotective drug to overcome or minimize gamma radiation induced biological damage in strategic armed personal, civilian population and cancer patients undergoing radiotherapy.

Biological Radiation Protection
A Protocol Manual

Dr. Raj Kumar

*Scientist-D and Head, Radiation Biotechnology Group,
Division of Radioprotective Drug Development and Research (DRDDR),
Institute of Nuclear Medicine and Allied Sciences (INMAS),
Defence Research and Development Organization (DRDO)*

2016

Daya Publishing House®

A Division of

Astral International Pvt. Ltd.
New Delhi – 110 002

Cataloging in Publication Data--DK
Courtesy: D.K. Agencies (P) Ltd. <docinfo@dkagencies.com>

Kumar, Raj *(Scientist-D),* **author.**
Biological radiation protection : a protocol manual / Dr. Raj Kumar.
   pages cm
Includes bibliographical references and index.
ISBN 978-93-86071-12-5 (International Edition)

   1. Radiation-protective agents. 2. Radiobiology. 3. Radiation-
-Physiological effect. I. Title.
RA1231.R2K86 2016   DDC 612.01448   23

Published by   : **Daya Publishing House®**
                 A Division of
                 Astral International Pvt. Ltd.
                 – ISO 9001:2015 Certified Company –
                 4736/23, Ansari Road, Darya Ganj
                 New Delhi-110 002
                 Ph. 011-43549197, 23278134
                 E-mail: info@astralint.com
                 Website: www.astralint.com

*I gratefully dedicated
this book to my Late Parents for
their impressive devotion and
hardship that make me
capable of authoring a book*

डॉ. अजय कुमार सिंह
उत्कृष्ट वैज्ञानिक/वैज्ञानिक 'एच'
निदेशक

Dr. A. K. Singh
Outstanding Scientist/Sc. 'H'
Director

अ.स.प.सं.INM/AKS//02/2016
DO No.
भारत सरकार, रक्षा मंत्रालय
GOVERNMENT OF INDIA, MINISTRY OF DEFENCE
रक्षा अनुसंधान तथा विकास संगठन
DEFENCE RESEARCH & DEVELOPMENT ORGANISATION
नाभिकीय औषधि तथा सम्बद्ध विज्ञान संस्थान
INSTITUTE OF NUCLEAR MEDICINE & ALLIED SCIENCES
ब्रिगे. एस. के. मजूमदार मार्ग, दिल्ली – 110 054
BRIG. S. K. MAZUMDAR MARG, DELHI - 110 054

Foreword

Despite tremendous efforts, medical management of nuclear radiation victims is still an area of challenges. Uses of nuclear radiation are increasing in the area of imaging, diagnostics, pharmaceutical research and cancer radiotherapy *etc.* Similarly, deliberate use of radiological dispersal device (RDD) by non state actors and even uses of nuclear weapon by nuclear power states cannot be ruled out. In any such scenario biological radioprotection will be an unavoidable necessity. However, more than five decades of intense research worldwide is unable to develop a non toxic radioprotective drug that can be used during any radiation emergency. Radiation induced oxidative stress in the cellular environment via inducing water radiolysis and free radicals generation, peroxidising lipids, oxidizing proteins, inducing DNA single and double strand's breaks lead to cell death. One of the most prominent causes of disqualification of the natural or synthetic radiation countermeasure agents was lack of systematic evaluation of their radioprotective efficacy and toxicity *in vitro*, *ex vivo* and *in vivo* models. Resulted, search for efficient non-toxic radioprotector is still continuing. Present systematic experimental manual to evaluate the efficacy of new radioprotective compounds at cellular, molecular and systemic levels will certainly provide excellent assistance to the radiation biologists. Author included optimized, tested and modified protocols in the manual to evaluate radioprotective efficacy of the test compound(s). This experimental manual will be beneficial to young researchers, scientists and academicians to evaluate radioprotective potential of new compounds in a comprehensive manner. I congratulate the author for his outstanding contribution in authoring this experimental protocol manual to facilitate research in the field of radiation biology and radioprotection.

(Dr. A.K. Singh)

अ.स.प.सं. / D.O. No. RKS/01//2016

भारत सरकार, रक्षा मंत्रालय
Government of India, Ministry of Defence

रक्षा अनुसंधान तथा विकास संगठन
Defence Research & Development Organisation

रक्षा खाद्य अनुसंधान प्रयोगशाला
Defence Food Research Laboratory

सिद्धार्थ नगर, मैसूरु / Siddarthnagar, Mysore - 570 011.

डॉ. राकेश कुमार शर्मा
निदेशक

Dr. Rakesh Kumar Sharma
Director

Foreword

Understanding of the biological effect of nuclear radiation and strategies to achieve biological radiation protection is always an area of tremendous challenge. It is astonishing fact that recognition of radiation damage to the biological tissue was an integral part of the discovery of radioactivity. Biological effect of nuclear radiation is a subject matter of radiation biology discipline that developed along with the nuclear physics and nuclear chemistry. With the increasing usage of nuclear technology or products like radioisotopes in diagnostics, therapeutic and research, and fission material in energy production, the chances of associated occupational risks and accidents are also rising. Besides that, live examples of deliberate nuclear weapon uses in Japan and nuclear reactor accidents in Chernobyl (in former USSR) and Fukushima, Japan warrants attention of scientists to develop efficacious and non-toxic radiation countermeasures. Despite five decades of consolidated efforts worldwide, search for ideal radioprotector is still continuing. Radiation induced oxidative stress in the living cells via inducing water radiolysis and free radicals generation, peroxidising lipids, oxidizing proteins, inducing DNA single and double strand's breaks results in disruption of cellular integrity alternately leading to cell death. One of the most prominent causes of non-approval of the natural or synthetic radiation countermeasure agents is by the regulating agencies lack of systematic evaluation of their radioprotective efficacy and toxicity *in vitro*, *ex vivo* and *in vivo* models. Present experimental protocol manual provides a systematic experimental basis of efficacy evaluation of radioprotective drug candidates at molecular , cellular and systemic levels. Author incorporated optimized and modified protocols to asses radioprotective efficacy of the test compound. I am sanguine that, the experimental manual will be equally beneficial to young researchers, scientists and academicians to evaluate radioprotective potential of premising compounds in a comprehensive manner. I compliment the author for his tremendous efforts to develop this hand book. I feel confident that users will find this manual as an intellectually enriching and unique book in the area of radiation biology and radiation protection. I once again congratulate the author for his outstanding contribution in authoring this experimental protocol manual to facilitate research in the important field of radiation biology and radioprotection .

(Dr. Rakesh Kumar Sharma)

Dr. Manju Gupta
Sc 'G, & Addl director
'Rakshak' Project Director
Head Division of RDDR
drmanjugupta2003@yahoo.com
TELEPHONE : 23905153/23946182
FAX No.:23919805
TELEGRAM : INMAS

Government of India
Ministry of Defence
Defence Research & Dev. Orgn.
Institute of Nuclear Medicine
& Allied Sciences (INMAS)
Brig SK. Mazumdar Road
Delhi – 110054, INDIA

Foreword

Biological system is highly sensitive to sudden change in high energy dissemination, and it responds to a great extent against radiation induced oxidative damage. However, beyond certain level, biological system is unable to combat radiation induced stress and thus require prophylactic and therapeutic interventions. Increasing use of nuclear energy for strategic and non strategic purposes proportionally increases the chances of radiation accidents. Additionally the use of nuclear weapon cannot be ruled out by enemy countries during a nuclear conflict at any point of time. In view of above, it is highly desirable to develop radiation countermeasures for prophylactic and therapeutic use. Despite several attempts to develop efficacious and non-toxic radioprotective drug, not a single safe agent has been approved for emergency situation/bed side use. Radiation induced oxidative damage occurs in the cells and tissues primarily by free radicals generated by radiolysis of water. Free radicals, immediately after generation, interact with macromolecules like DNA, proteins and lipids. Simultaneously oxidation of endogenous enzymes leads to the impairment of cellular functionality. Therefore, protection of cell and tissue integrity against radiation is a necessity to preserve biological system. The agents that are able to minimize radiation induced immune-suppression and apoptosis inhibition can be the potential candidates for radioprotection. Present experimental protocol manual will provide an insight for systematic evaluation of radiation countermeasure agents. I compliment the author for his extensive efforts to prepare this systematic experimental manual and strongly feel that the researchers working in the field of radiation countermeasures will be greatly benefited by this. I personally congratulate the author for his outstanding contribution to facilitate research in the field of radiation biology and radioprotection.

(Dr. Manju Gupta)

Preface

Development of efficacious nontoxic radioprotective drug is an un-cherished dream of radiation biologist throughout the world. Despite several attempts, by enlarge either no successes or limited successes are achieved and search of clinically accepted radioprotective agents still continued. The importance and necessity of writing the experimental protocol manual to screen out and evaluate novel radioprotective drug was realised because large number of researchers working in the area of biological radioprotection worldwide using random methodology for radioprotector screening. No systematic experimental manual is available which can be referred to evaluate the radioprotective efficacy of a natural or synthetic agent systematically. Gamma radiation induced oxidative damage to the biological environment at atomic, molecular, cellular and systemic level with varying degree of sensitivity. For example, *in vitro* condition radiation induced proteins damage (carbonylation) is visible only beyond 500 Gy of radiation doses, while, *in vivo* conditions similar level of protein carbonylation can be achieved at 5-10 Gy of gamma radiation in mice model. Therefore, to measure the radiation induced protein damage, common experimental protocol cannot be used for *in vitro* as well as *in vivo* conditions. Similarly, radiation induced oxidative damage to DNA *in vitro* and *in vivo* situations represent a wide gap of radiation doses need to be used in experimental setting. For example, *in vivo* models, 2-5 Gy gamma radiation doses are more than enough to observed significant single or double strand break in genomic DNA. However, similar level of DNA damage *in vitro* condition required at least 200Gy of gamma radiation doses. Since, gamma radiation induced oxidative stress in the biological tissues primarily by the radiolysis of water and generation of free radicals. Thus, evaluation of *in vitro* and *in vivo* free radicals scavenging potential of the radioprotective drug is an immediate necessity to categories the class of the radioprotective drug as antioxidant. Therefore, a full section **(Section I)** was presented in the present manual to estimate the preliminary *in vitro* antioxidant potential of the radioprotective compound under evaluation.

Further, to assess *in vitro* radiation induced oxidative damage to the bio-molecules (*i.e.* lipid, protein and DNA) and protecting capability of radioprotective drug, a complete section **(Section II)** of secondary methods of radioprotection was incorporated in the manual. To determine the *in vitro* radioprotective efficacy of the test compound(s) using cell culture models, CFU estimation, SOD, catalyse, glutathione reductase, glutathione-S-transferase, lysosomal membrane stability and total ROS generation etc., estimation protocols with reference to gamma radiation damage was placed **(Section III)** under the name of advance methods for radioprotection.

Gamma radiation induced DNA damage and apoptosis *in vitro* as well as *in vivo* conditions. Therefore, to measure radioprotection offered by radioprotective drug to radiation induced DNA damage and apoptosis, a complete section **(Section IV)** was included in the manual. Gamma radiation severely perturbs the mitochondrial DNA, resulted impaired mitochondrial functions lead to energy depletion and ultimately cell death. Therefore, to achieve efficient radioprotection, repair of radiation induced mitochondrial DNA damage is unavoidable necessity. To evaluate mitochondrial DNA radioprotection/repair capability of a radioprotective compound a dedicated section **(Section V)** was included in the manual. This section comprises with the protocols associated with DNA repair like DNA glycosylase (OGG1), UDG, AP endonuclease, DNA polymerase g gap filling activity, 8-oxoGmtDNA adduct detection and histone deacetylase activity determination.

Mitochondria are highly sensitive to radiation induced oxidative damages. Being a respiratory centre, mitochondria always carry high oxidative stress. Under gamma irradiation, redox homeostasis of the organelle disturbs resulted ATP synthesis compromised. Therefore, radioprotection of mitochondria needs attention. We search across the literature, but negligible studies are available which focused mitochondrial oxidative phosphorylation in radioprotection. Being highly ignored, no specific literature about the experimental protocols to study mitochondrial oxidative phosphorylation is available. Therefore, in the present manual, some specific assay associated with oxidative phosphorylation **(Section VI)** *i.e.* study of respiratory complex I, II, III, IV and IV, or cytochrome C release and ATPase assays, mitochondrial outer membrane permeabilization potential and mitochondrial total antioxidant status determination assay etc. were specifically modified in terms of radioprotection efficacy demonstration of a radioprotective drug at mitochondrial level.

Among all, immune system is highly radiosensitive system. Bone marrow stem cells which continuously form various haematopoietic cells are always in the states of proliferation and thus prone to radiation mediated DNA damage. Followed by irradiation, lymphocytes counts seriously deplete and use as indicator of radiation exposure. Furthermore, due to radiation induced mortality in bone marrow stem cells, overall cells of haematopoietic lineage gets suppress lead to immune system impairment. Therefore, a radioprotective drug must take care of haematopoietic system without that radioprotection is an impossible task. Thus, to evaluate radioprotective potential of a drug, its effect on haematopoietic growth factors *i.e.* G-CSF, GM-CSF, M-CSF, chemokine and cytokine expression, NFkB activation,

Th1/Th2 cytokine homeostatsis evaluation, T cells, B cells and monocytes/ macrophages cells population should be determination. In view of above, **Section VII** was formulated to determine radioprotective action of a radioprotective drug towards immune system.

It is expected that the information given in this experimental manual may prove to be beneficial to the scientists, clinicians and research students working in the area of radioprotection, radiation biology and radiotherapy.

I sincerely acknowledge **Dr. R.P. Tripathi,** Outstanding Scientist and Director INMAS for his full-hearted support and guidance to make this experimental protocol manual as reality. I am feeling proud to acknowledges **Dr. HC Goel,** Scientist G (Retd.) INMAS, DRDO, currently positioned as Director, Amity Centre for Radiation Biology, Amity University, Noida, UP, who introduce me to the subject of Radiation Biology and Radioprotection as my first mentor at INMAS in the year of 2001. I extend my deep gratitude to **Dr. R.K. Sharma,** Scientist G, HOD CBRN Division, INMAS for his valuable suggestions, guidance and appreciation before and during formulation of this manual. I also extent my thankfulness to **Dr. Manju Gupta,** Scientist G, HOD Division of Radioprotective Drug Development and Research (DRDDR), INMAS for her whole hearted support guidance and appreciation during realization of this experimental protocol manual.

I also express my deep gratitude to all fellow scientists at INMAS and outside INMAS for their full support and motivation to complete the book. I also extend honest thanks to my all personal and especial friends who have given me constant support and encouragements. Last but not the least, I would like to acknowledge my family members particularly my wife Rupa Khodlan, daughter Pallavi Khodlan and son Himanshu Khodlan for their patient and dedication to give me all possible support to complete this book.

Raj Kumar

Contents

Section I
Preliminary *In vitro* Methods to Screen Out Radioprotective Compounds

Section IV
Determination of Radiation Induced Apoptosis and DNA Damage

Section V
Radiation Induced Mitochondrial DNA Damage and its Modulation by Radioprotective Drug Treatment

Section VI
Radiation Induced Mitochondrial Oxidative Phosphorylation Perturbations and their Modulation by Radioprotective Compounds Treatment

Section VII
Immune System Radioprotective Efficacy Evaluation of Radioprotective Compounds

Section 1

Preliminary *In vitro* Methods to Screen Out Radioprotective Compounds

Absorption/exposure of gamma radiation energy by biomolecules in chemical or biological environment, lead to transform them in to highly reactive radical species (*i.e.* ROS/RNS). In cellular environment, water radiolysis due to gamma radiation exposure, generate short live OH* and H* radicals and thus lead to cellular oxidative stress. Various biomolecules including structural and functional proteins, membrane lipids and genetic material *i.e.* DNA get oxidized by free radicals cascading (Fenton) reaction resulted cellular, biochemical and biomolecular functions impairement. Therefore, free radicals neutralizing strategy seems to be instrumental to either minimize/completely inhibit the radiation induced damage in cellular milieu. In view of above, screening of potential antioxidant molecules/agents able to scavenge free radicals in irradiated environment is of utmost significance. Thus, following (Figure 1) free radicals scavenging assays to screen out excellent antioxidant/free radicals scavengers are summarized in next pages.

Protocol No. 1.1: Colorimetric Estimation of Reducing Capability of a Radioprotective (Antioxidant) Drug using Stable DPPH Radicals

Principle of the Assay

DPPH commonly abbreviated for 2,2-diphenyl-1-picrylhydrazyl a well known stable free-radical molecule. Any antioxidant molecule able to donate its electron to DPPH resulted neutralization of the DPPH radicals lead to discoloration and decrease absorbance at 520nm. Thus, electron donation potential of an antioxidant molecule will be directly proportional to decrease absorbance of the DPPH. That can be plotted and estimated in terms of IC_{50} value. The IC_{50} values for ascorbic acid and propyl gallate were calculated as 11.8 µM and 4.4 µM in methanol respectively. Gamma radiation is known to induced free radicals via radiolysis of water in biological system. Therefore, a chemical compound able to neutralize free radical such as DPPH in vitro can be act as radioprotector (Figure 2). Thus the DPPH assay has direct relevance to radioprotector screening in vitro conditions (1-6).

Assay Requirement

Microtubes (2.0 ml), micro-pipettes (volume range 200-1000µl), DPPH, methanol, UV-Vis spectrophotometer.

1.1 Assay Procedure

1.1.1 DPPH Standard Curve Preparation

1.1.1.1 A stock (80µM) solution of DPPH was prepared in methanol.

1.1.1.2 A series of different dilutions (range from10-1000µl) from DPPH stock solution was prepared.

1.1.1.3 Final volume of the reaction mixture was make-up to 1.1 ml with methanol.

1.1.1.4 The absorbance of the reaction mixture was recorded at 517nm at room temperature.

1.1.1.5 A standard curve between absorbance and concentration of DPPH was plotted.

1.1.2 DPPH Radical Neurilizing Activity of Test Compound

1.1.2.1 A stock solution (1.0mg/ml) of the compound (antioxidant/radioprotector) under evaluation was prepared preferably in doubled distilled water or alcohol.

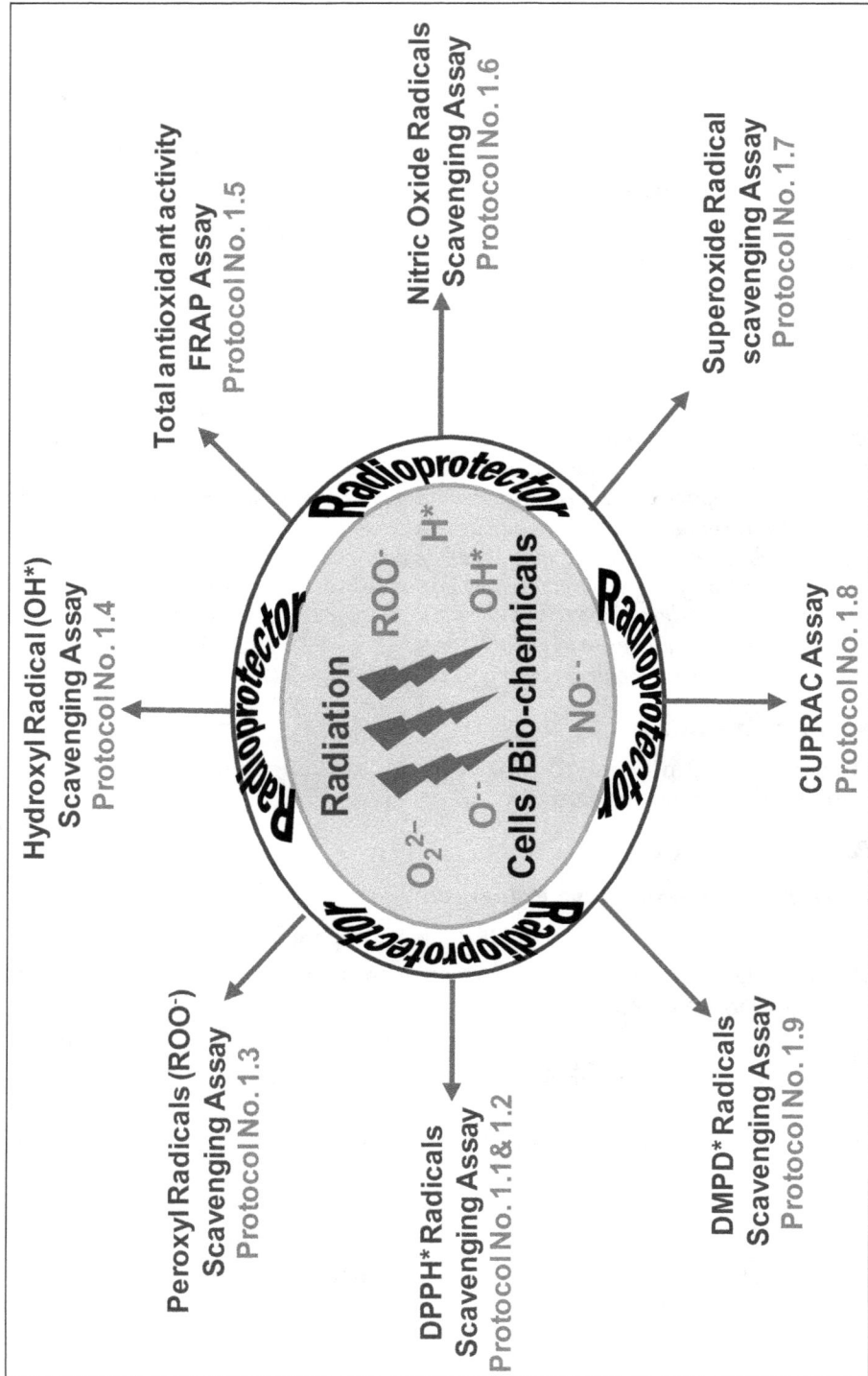

Figure 1: Different Kinds of Antioxidant Assays to Screen Out Free Radicals Scavenging Agent(s).

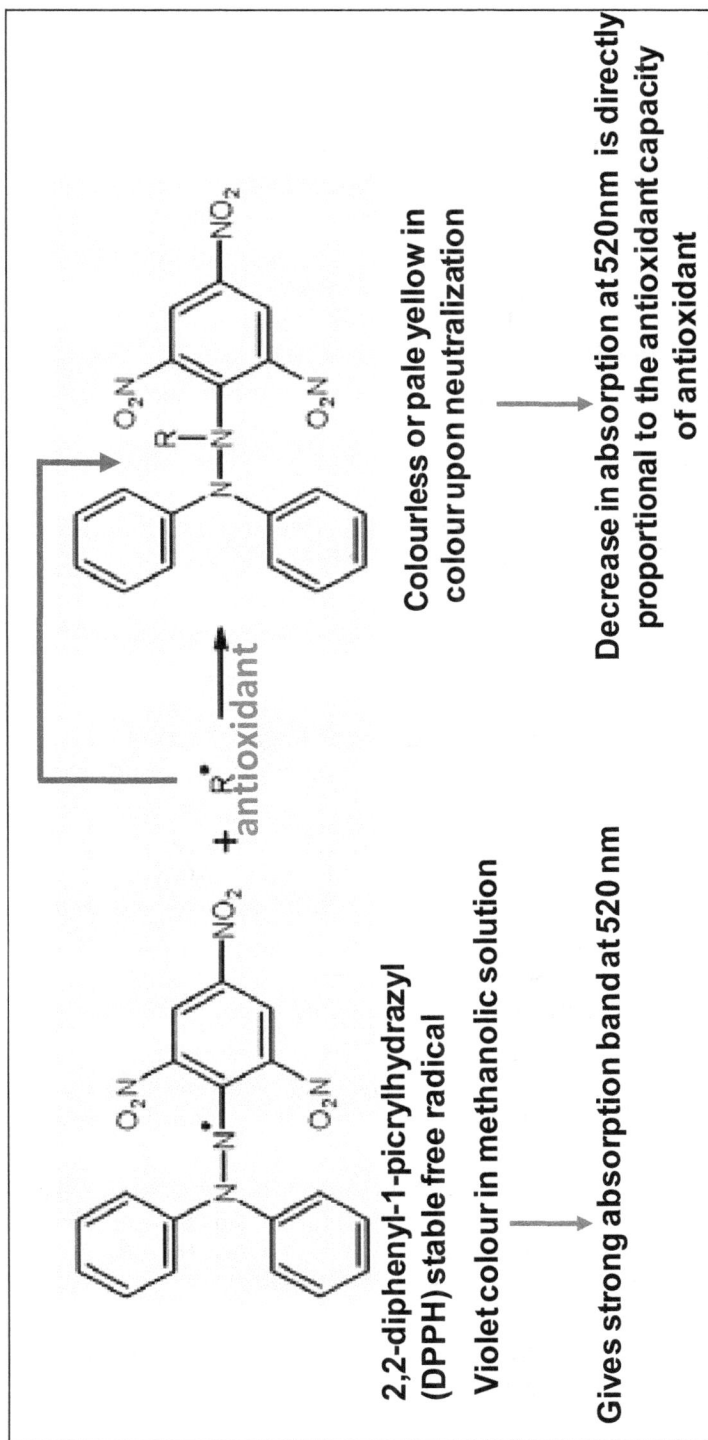

2,2-diphenyl-1-picrylhydrazyl (DPPH) stable free radical

Violet colour in methanolic solution

Gives strong absorption band at 520 nm

Colourless or pale yellow in colour upon neutralization

Decrease in absorption at 520nm is directly proportional to the antioxidant capacity of antioxidant

Figure 2: Mechanism of DPPH Neuterilization by an Antioxidant Compound.

1.1.2.2 Series of different dilutions (10-100µl) of test compound was prepared in 2.0 ml tubes and final volume makeup 1.0 ml with doubled distilled water or alcohol.

1.1.2.3 To ensure the proper mixing of the compound with diluent, tubes should be stirred at least for 5.0 minutes with the help of stirrer at room temperature.

1.1.2.4 1.0 ml of DPPH (80µM) was added into each tube in the dark.

1.1.2.2 The mixture was incubated at room temperature for 10 minutes in the dark.

1.1.2.5 The negative control was prepared simultaneously by replacing test compound with distilled water/alcohol.

1.1.2.6 Absorbance of the reaction mixture was recorded at 520nm at room temperature.

1.1.2.7 Percent reduction in the absorbance (OD) of DPPH radicals was calculated using formula:

Absorbance of TEST compound/absorbance of the control x 100

1.1.2.8 IC_{50} value of neutralizing DPPH radicals for individual compound was calculated using standard curve of DPPH prepared in the similar *in vitro* conditions.

References

1. Szabo M R, Idiþoiu C, Chambreand D and Lupea A X (2007). Improved DPPH determination for antioxidant activity spectrophotometric assay. *Chemical Papers*. 61(3), 214-216.

2. Thaipong K, Boonprakob U, Crosby K, Cisneros-Zevallos L and Byrne DH (2006). Comparison of ABTS, DPPH, FRAP, and ORAC assays for estimating, antioxidant activity from guava fruit extracts. *J. Food Composi. Analy.* 19, 669–675.

3. Apak R, Gorinstein S, Böhm V, Schaich K M, Özyürek M, and Güçlü K (2013). Methods of measurement and evaluation of natural antioxidant capacity/ activity (IUPAC Technical Report). *Pure Appl. Chem.* 85(5), 957–998.

4. Meléndez N P, Nevárez-Moorillón V, Rodríguez-Herrera R. Espinoza J C and Aguilar C N (2014). A microassay for quantification of 2,2-diphenyl-1-picrylhydracyl (DPPH) free radical scavenging. *African J. Biochem. Res.* 8(1), 14-18.

5. Kumar R, Bansal D D, Patel D D, Mishra S, Karamalakova Y, Zheleva A, Gadjeva V and Sharma R K (2011). Antioxidative and radioprotective activities of semiquinone glucoside derivative (SQGD) isolated from *Bacillus* sp. INM-1. *Mol. Cell. Biochem.* DOI 10. 1007/s11010-010-0660-x.

6. Zheleva A, Karmalakova Y, Nikolova G, Kumar R, Sharma R and Gadjeva V. (2011). A new antioxidant with natural origin characterized by electron paramagnetic resonance spectroscopy methods. *Pharmaceu. Biotech.* DOI: 10.5504/50YRTIMB. 2011.0027.

Protocol No. 1.2: Anti-Radical Power (ARP) Estimation using DPPH Radicals Neutralizing (IC$_{50}$) Activity of a Radioprotective (Antioxidant) Agent in Terms of Molar Concentration

Assay Principle

Please refer protocol No. 1.1 (Page No. 3). The only difference in this assay from previous assay was that here ARP was expressed as the number of moles of antioxidant/mole DPPH* instead of per cent reduction in DPPH absorbance as taken in account in previous assay. To perform this assay it is mendatory that antioxidant compound under evaluation should be in pure form with define molecular weight. Present assay is not applicable for the samples have mixture of compound like herbal extract etc.

Assay Requirement

Microtubes (2.0 ml), micro-pipettes (volume range 200-1000µl), DPPH, methanol, UV-Vis spectrophotometer.

1.2.1 Assay Procedure

1.2.1.1 Different concentrations (mM) of pure antioxidant molecule (pure molecule with defined molecular weight) were prepared in distilled water or alcohol.

1.2.1.2 100µl of antioxidant solution was added to 3.9 ml of DPPH radical solution (stock solution 6×10^{-5}mol/l).

1.2.1.3 The decrease in absorbance was determined at 515-520nm at 0, 1 minutes and repeated at every 15 minutes until the reaction reached to saturation plateau.

1.2.1.4 The exact initial DPPH* concentration (C_{DPPH}) in the reaction mixture was calculated from a calibration curve with the help of equation:

Abs at 520nm = $12,509 \times (C_{DPPH})$-2.58×10^{-3}as determined by linear regression. Reaction kinetics for antioxidant compound was plotted.

1.2.1.5 The molar concentration of DPPH* radicals corresponding to the steady state was determined in percentage of DPPH radicals remaining in the mixture.

1.2.1.6 The values transferred onto another graph showing the percentage of residual DPPH* radicals in the reaction mixture at the steady state as a function of the molar ratio of antioxidant to DPPH.

1.2.1.7 Antiradical activity of particular compound was defined as the amount of antioxidant compound required to decrease the initial DPPH*radicals concentration by 50 per cent.

1.2.1.8 For reasons of clarity, we will speak in terms of IC_{50} or the antiradical power (ARP). An increase in ARP means the more efficient antioxidant (1-7).

References

1. Brand-Williams W, Cuvelier M E and Berset C. (1995).Use of a Free Radical Method to Evaluate Antioxidant Activity. *Lebensm.-Wiss Technol.*, 28, 25-30.

2. Szabo M R, Idiþoiu C, Chambreand D and Lupea A X (2007). Improved DPPH determination for antioxidant activity spectrophotometric assay. *Chemical Papers.* 61(3), 214-216.

3. Thaipong K, Boonprakob U, Crosby K, Cisneros-Zevallos L and Byrne DH (2006). Comparison of ABTS, DPPH, FRAP, and ORAC assays for estimating, antioxidant activity from guava fruit extracts. *J. Food Composi. Analy.* 19, 669–675.

4. Apak R, Gorinstein S, Böhm V, Schaich K M, Özyürek M, and Güçlü K (2013). Methods of measurement and evaluation of natural antioxidant capacity/activity (IUPAC Technical Report). *Pure Appl. Chem.* 85(5), 957–998.

5. Meléndez N P, Nevárez-Moorillón V, Rodríguez-Herrera R. Espinoza J C and Aguilar C N (2014). A microassay for quantification of 2,2-diphenyl-1-picrylhydracyl (DPPH) free radical scavenging. *African J. Biochem. Res.* 8(1), 14-18.

6. Kumar R, Bansal D D, Patel D D, Mishra S, Karamalakova Y, Zheleva A, Gadjeva V and Sharma R K (2011). Antioxidative and radioprotective activities of semiquinone glucoside derivative (SQGD) isolated from *Bacillus* sp. INM-1. *Molecular and Cellular Biochemistry.* DOI 10. 1007/s11010-010-0660-x.

7. Zheleva A, Karmalakova Y, Nikolova G, Kumar R, Sharma R and Gadjeva V. (2011). A new antioxidant with natural origin characterized by electron paramagnetic resonance spectroscopy methods. *Pharmaceutical Biotech.* DOI: 10.5504/50YRTIMB. 2011.0027.

Protocol No. 1.3: Gamma Radiation or Chemical Oxidant Induced Peroxyl (ROO⁻) Radicals and their Neutralization Estimation using Oxygen Radical Absorbance Capacity (ORAC) Assay

Assay Principle

The ORAC assay worked upon the principle of neutralization of the peroxyl (ROO⁻) radicals which were induced either by oxidation started by 2,2'-Azobis (2-amidinopropane) dihydrochloride (AAPH) in temperature (37°C) depended decomposition or by gamma radiation in chemical or biological environment. The ROO⁻ radicals react with fluorescein and oxidize it instatntly, lead to flourescein conversion into a non fluorescence product that can be observed using fluorescence spectrometry using excitation 488nm and emission 535nm. Antioxidant molecules can slow down, suppress or completely inhibits this reaction by hydrogen atom donation to the oxidize fluorescein. The fluorescence signal decaying rate in presence and absence of antioxidant molecule can be measured using time lasped (30 minutes) fluorescence spectrometric analysis. The exact concentration of the antioxidant molecules in the test sample will be directly proportional to the fluorescein's fluorescence intensity (1-3). That can be assessed by comparing the total area under the curve compared to a known antioxidant, i.e.Trolox. A schematic representation of basic principle was placed (Figure 3) below.

Assay Requirement

2,2'-Azobis(2-amidinopropane) dihydrochloride (AAPH), phosphate buffer (pH 7.4), fluorescein, trolox and fluorescence spectrometer.

1.3.1 Assay Procedure

1.3.1.1 2,2'-Azobis(2-amidinopropane) dihydrochloride (AAPH; *75 mM*) fresh solution was prepared by dissolving 203.4 mg AAPH in 10 ml of phosphate buffer (75mM) saline (pH 7.4).

1.3.1.2 A fresh 4 µM fluorescein stock solution was prepared with 75 mM phosphate buffer (pH 7.4) and stored at 4 °C in the dark to avoid photo-bleaching.

1.3.1.3 Just before to use, the fluorescein stock solution was diluted upto 1:500 with PBS (pH 7.4).

1.3.1.4 The diluted sodium fluorescein working solution should be prepared freshly just before the experiments.

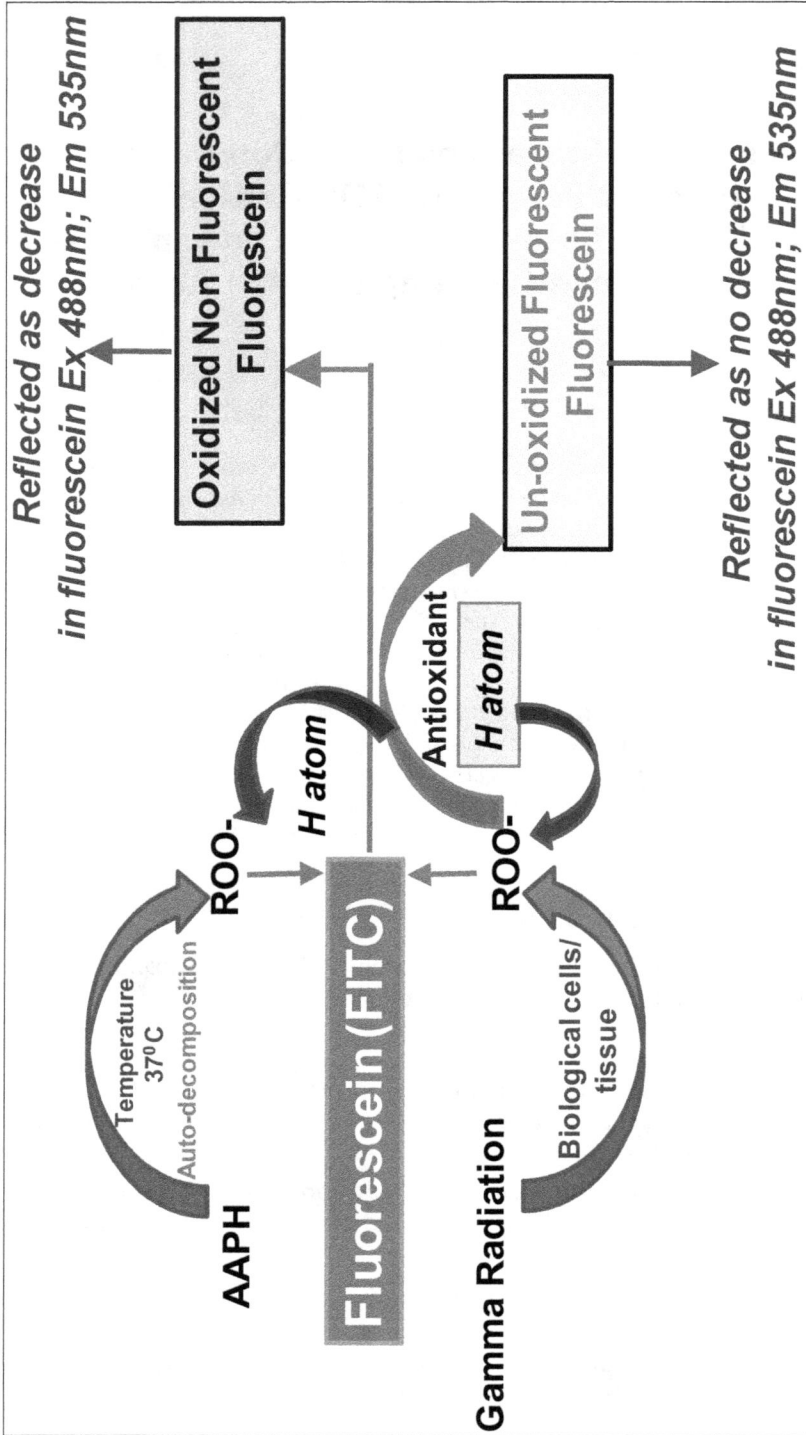

Figure 3: Mechanism of Peroxyl Radicals Generation and their Neutralization by an Antioxidant Agent.

1.3.1.5 Sodium fluorescein (150 μl) working solution was added to the desired wells of 96 well plates.

1.3.1.6 While different concentrations of standards Trolox solution 25 μl or test sample 25 μl was added to the wells.

1.3.1.7 The test plate (containg following experimental groups:

   (i) AAPH/IR+fluorescein+ trolox

  (ii) AAPH/IR+Fluorescein+test samples and

 (iii) AAPH+Fluorescein–trolox/test sample

       Reaction mixture was then allowed to equilibrate by incubating for a minimum of 30 minutes at 37 °C.

1.3.1.8 The fluorescence decay of oxidize fluorescein in absence of trolox or test sample (concentrations depended manner) was then monitored kinetically with absorbance (ex 488; em 535nm) recording continuously for 20 minutes at one minutes intervals.

1.3.1.9 Blank wells was simultaneously prepared by replacing fluorescein with 25 μl of 75 mM phosphate buffer (pH 7.4).

References

1. Rodrigues E, Mariutti LR, Chisté RC, Mercadante AZ (2012). Development of a novel micro-assay for evaluation of peroxyl radical scavenger capacity: application to carotenoids and structure-activity relationship. *Food Chem.* 135(3): 2103-2111.

2. Lo Scalzo, Todaro and Rapisarda (2012). Methods used to evaluate the peroxyl (ROO.) radical scavenging capacities of four common antioxidants. *European Food Res. Technol.* DOI: 10.1007/s00217.

3. Valkonen M and Kuusi T. (1997). Spectrophotometric assay for total peroxyl radical trapping antioxidant potential in human serum. *J Lipid Res.* 38: 823-833.

Protocol No. 1.4: Colorimetric Estimation of Hydroxyl Radical (OH*) Scavenging Activity of a Radioprotective Compound using Deoxyribose Degradation Assay

Assay Principle

The hydroxyl radical scavenging activity of an antioxidant agent can be measured calorimetrically by studying the competition between deoxyribose and antioxidant molecule for scavenging of hydroxyl radicals generated with Fe^{3+}-ascorbate- EDTA-/H_2O_2 system. The hydroxyl radicals generated by chemical systems instantly attack deoxyribose that lead to TBARS formation. TBARS formed can be quantified spectrophotometrically (1-6). A schematic representation (Figure 4) of the chemical regulation of TBARS formation and its antioxidant mediated inhibition is shown below:

Assay Requirement

Microtubes (2.0 ml), micro-pipettes (200-1000µl), ferric chloride, L-ascorbic acid, deoxyribose sugar potassium phosphate buffer (pH 7.4), H_2O_2 Trichloroacetic acid (TCA), thiobarbituric acid (TBA), NaOH, UV-Vis spectrophotometer.

1.4.1 Assay Procedure

The hydroxyl radicals scavenging potential of an antioxidant/radioprotector can be measured using the deoxyribose degradation assay.

1.4.1.1 A fixed volume (500 µl) of different concentration of radioprotective compound was mixed with 500µl of assay reagent containing ferric chloride (100 µM) solution, L-ascorbic acid (100µM along with H_2O_2 (1mM) and deoxyribose (3.6 mM) in potassium phosphate buffer (pH 7.4) in a total assay volume of 1.0 ml.

1.4.1.2 Reaction mixture was incubation for 1h at room temperature.

1.4.1.3 1.0 ml of TCA (10 per cent w/v) and 1.0 ml of thiobarbituric acid (0.5 per cent w/v, TBA in 0.025 M NaOH) was added to each sample tube and reincubated in a hot water bath at 55°C for 15 minutes.

1.4.1.4 The tubes were kept for cooling at room temperature for 30 minutes.

1.4.1.5 Absorbance was recorded at 532 nm against the blank.

1.4.1.6 Blank sample was prepared simultaneously in a similar way but without drug content.

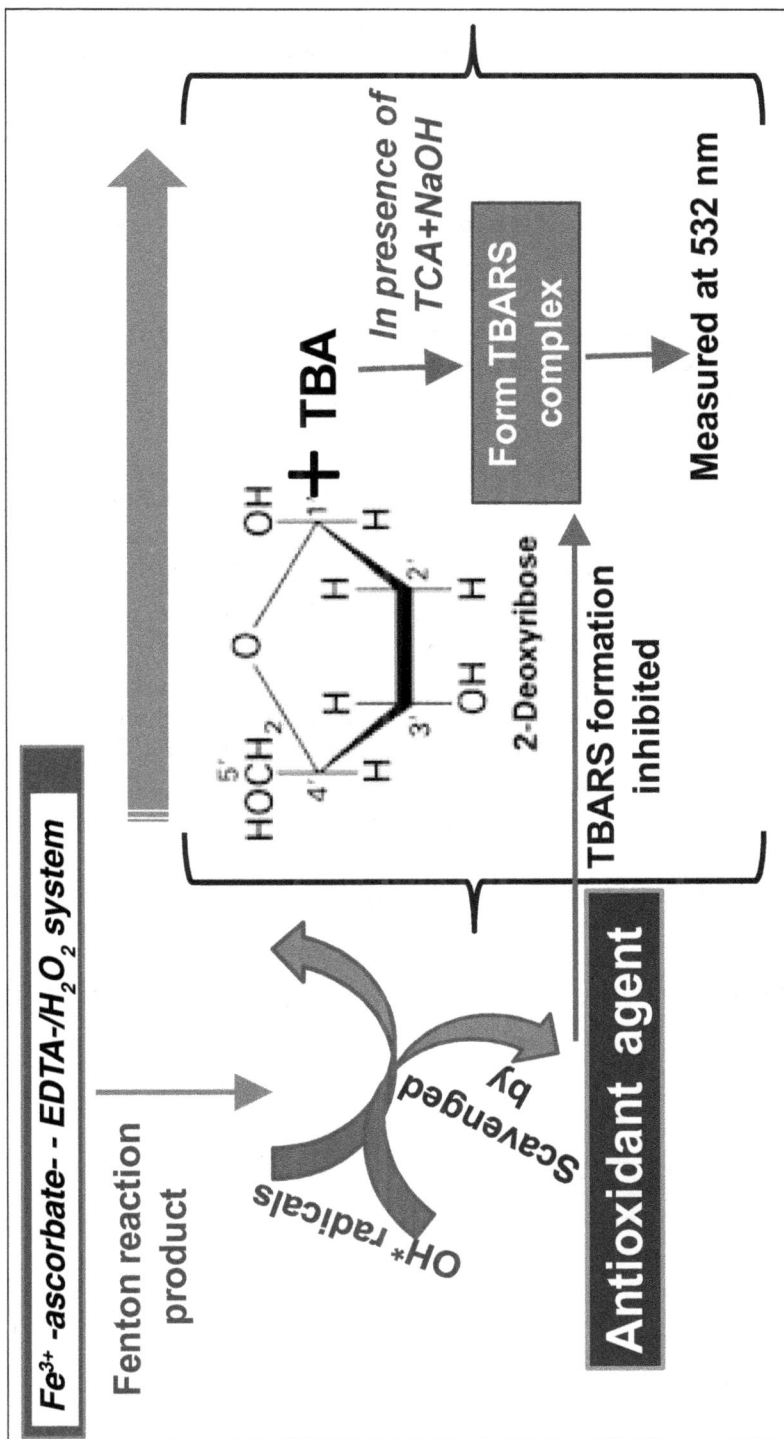

Figure 4: Schematic Representation of Chemical Mechanism of OH· Radicals Scavenging and TBARS Complex Formation inhibition by Antioxidant Agent.

TBA: Thiobarbutaric acid, TCA: Trichloroacetic acid.

1.4.1.7 The decrease in the absorbance with increasing drug concentrations was directly proportional to the increasing hydroxyl radical scavenging potential of the test compound.

1.4.1.8 The percentage (per cent) inhibition of deoxyribose degradation was calculated in terms of hydroxyl radical scavenging potential of a test compound using following formula:

Per cent Inhibition = (O.D. control - O.D.sample/O.D. control) × 100

References

1. Chandran RP, Vysakhi, MV, Manju S, Kannan M, Abdul Kader S and Sreekumaran Nair, A (2013). *In vitro* free radical scavenging activity of aqueous and methanolic leaf extracts of A*egle tamilnadensis* abdul kader (rutaceae). *Inter. J. Pharm. Pharmaceut. Sci.* 5(3), 819-823.

2. Chen S and Schopfer P (1999). Hydroxyl-radical production in physiological reactions A novel function of peroxides. *Eur. J. Biochem.* 260, 726-735.

3. Li X (2013). Solvent effects and improvements in the deoxyribose degradation assay for hydroxyl radical-scavenging. *Food Chem.* 141(3), 2083-2088.

4. Hazra B, Biswas S and Mandal N (2008). Antioxidant and free radical scavenging activity of *Spondias pinnata. BMC Complem. Alter. Med.* 8, 63-73.

5. Cheng Z, Li Y and Cheng W (2003). Kinetic deoxyribose degradation assay and its application in assessing the antioxidant activities of phenolic compounds in a Fenton reaction system. *Analytica Chimica Acta.* 478, 129-137.

6. Pal R, Girhepunje K, Shrivastav N, Hussain M M and Thirumoorthy N (2011). Antioxidant and free radical scavenging activity of ethanolic extract of *Morinda citrifolia. Annal. Biol. Res.* 2(1), 127-131.

7. Ghosh A K, Mitra E, Dutta M, Mukherjee D, Anjali Basu A, Firdaus SB, Bandyopadhyay D, Chattopadhyay A (2013). Protective effect of aqueous bark extract of *Terminalia arjuna* on Cu^{2+}-ascorbate induced oxidative stress *in vitro*: involvement of antioxidant mechanism(s). *Asian J. Pharm. Clin. Res.* 6, (1), 196-200.

Protocol No. 1.5: Colorimetric Estimation of Reducing Capacity (*i.e.* total antioxidant activity) of a Radioprotective Compound

Assay Principal

This assay involves measuring conversion of a Fe^{3+} or ferricyanide complex to the ferrous cyanide complex form. The antioxidant agent under investigation easily donate an electron to the potassium ferriccyanide complex and convert ferric iron in to ferrous iron. The rate and degree of conversion of Fe^{3+} (ferric form) in to Fe^{2+} form after electron acceptance lead to complex colour chages from yellowish brown to bluish colour that can be measured spectrophometrically at 560nm (1-4). The chemical reaction represented (Figure 5) the principle of the the assay mentioned below:

Assay Requirement

Phosphate buffer pH 6.5, Potassium ferricyanide (K_3FeCN_6), Trichloroacetic acid (TCA), Ferric chloride ($FeCl_3$), Microtubes (2.0ml), Centrifuge, Spectrophotometer, Water bath, Incubator and Micropipettes (200-1000ml).

1.5.1 Assay Procedure

The reducing power of an antioxidant compound (total antioxidant activity) can be determined according to the method described below:

1.5.1.1 Reaction mixture prepared was contained 200µl of 0.2 M phosphate buffer (pH 6.5), 50µl of drug (Drug conc. varying from 1-2000µg/ml) and 200µl of K_3FeCN_6 in 2.0 ml test tube.

1.5.1.1 Reaction mixture was then incubated at 50°C for 20 minutes.

1.5.1.2 200µl of 10 per cent TCA was added in each tube and again incubated for 10 minutes at room temperature.

1.5.1.3 500µl of double distilled water and 100µl of 0.1 per cent $FeCl_3$ was added in each sample tube.

1.5.1.4 Finally, the samples were incubated at 37°C for 10 minutes and absorbance (O.D.) was recorded at 700nm.

1.5.1.5 The electron donation potential (*i.e.* reducing power) of radioprotective compound was calculated on the basis of their respective concentrations (mg/ml) using the formula:

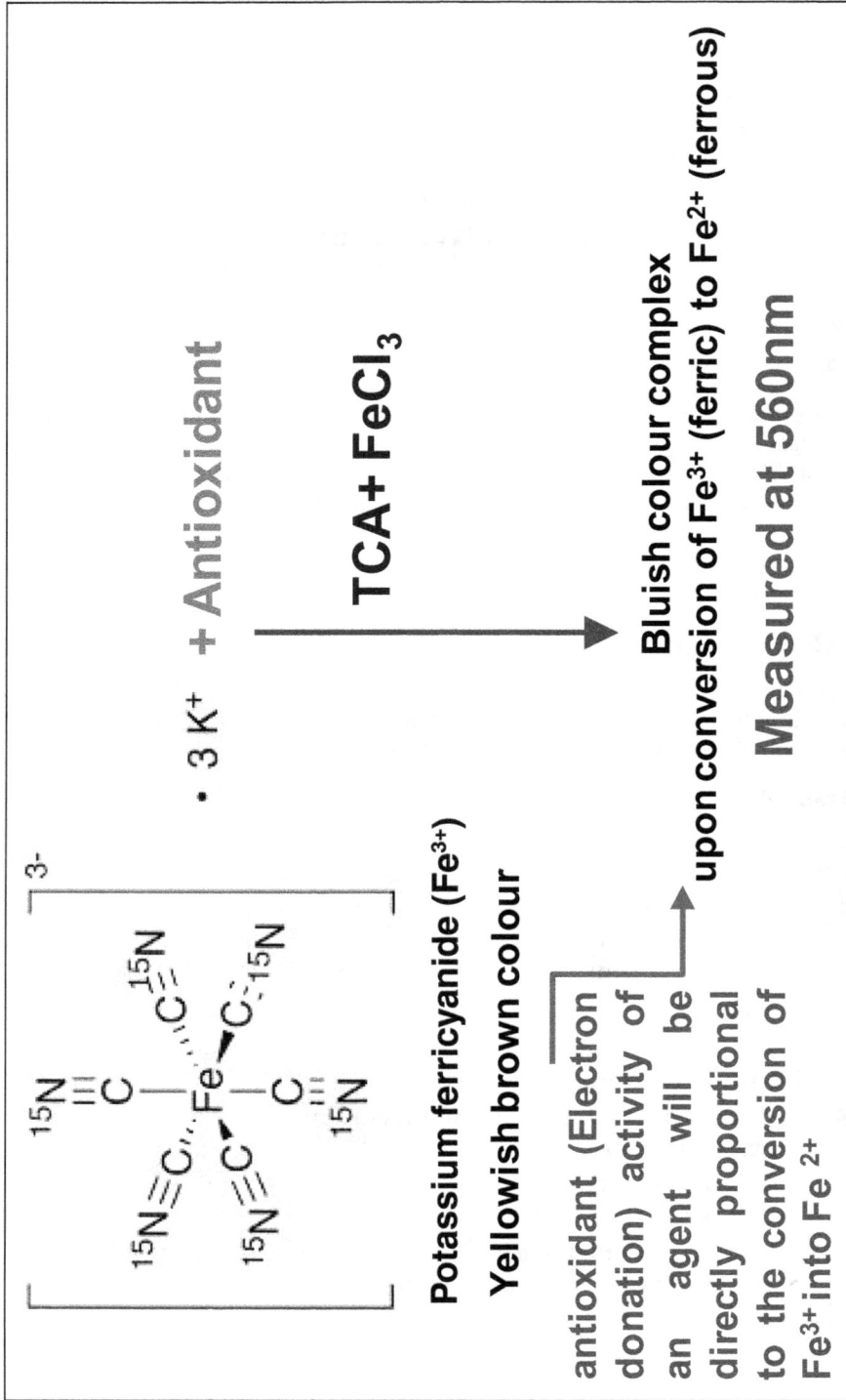

Figure 5: Schematic Representation of FRAP Assay Principle and Electron Donation/Acceptance Process Controlled by Antioxidant Agent and Oxidant Reactant.

Concentration (mg/ml) unit abs. value=$C_1/Abs.C_1+C2/Abs.C_2+C_3/Abs.C_3$

Where, C_1, C_2 and C_3 are representative three randomly selected concentrations (mg/ml) from the linear response curve. Increased absorbance was indicative of increased electron donation potential of a drug compound.

References

1. Benzie I F F and Strain J J (1996). The Ferric Reducing Ability of Plasma (FRAP) as a Measure of "Antioxidant Power": The FRAP Assay. *Anal. Biochem*, 239 (1): 70-76.

2. Huang D, Ou B and Prior R L (2005). The Chemistry behind Antioxidant Capacity Assays. *J. Agric. Food Chem*. 53, 1841-1856.

3. Thaipong K, Boonprakob U, Crosby K, Cisneros-Zevallos L, Byrne D H (2006).Comparison of ABTS, DPPH, FRAP, and ORAC assays for estimating, antioxidant activity from guava fruit extracts.*J. Food Composi. Analy.* 19 (2006) 669–675.

4. Szeto Y T, Tomlinson B and Benzie I F F (2002). Total antioxidant and ascorbic acid content of fresh fruits and vegetables: implications for dietary planning and food preservation. *Br. J. Nutrition*. 87, 55–59.

Protocol No. 1.6: Colorimetric Estimation of Nitric Oxide Radicals Scavenging Properties of a Radioprotective Compound

Assay Principle

Sodium nitroprusside in aqueous solution at physiological pH7.4 spontaneously generates nitric oxide, which interacts with oxygen to produce nitrite ion. Upon sulphonic acid addition, the nitrite forms adiazonium complex. After that when (N-alpha-naphthylethylenediamine was added, a pink coloured complex was formed (Figure 6) that can be measured at 540nm spectrophotometrically (1-5).

Assay Requirement

Sodium nitroprusside, Griess reagent, sulphanilamide, Hydrochloric acid,napthylethylenediamine hydrochloride, orthophosphoric acid, Centrifuge, Spectrophotometer, Water bath, Incubator, Micro-centrifuge tube and Micropipettes.

1.6.1 Assay Procedure

Nitric oxide radicals scavenging activity of test compound can be determined according to the method described below:

1.6.1.1 Varied concentrations of the test compound were mixed with fix volume of sodium nitroprusside (10 mM) and the reaction volume made up to 1.0 ml using phosphate buffer saline (pH 7.4).

1.6.1.2 1.0 ml of Griess reagent (composition: 6 per cent sulphanilamide in 3N HCl, 0.3 per cent napthylethylenediamine hydrochloride, 7.5 per cent orthophosphoric acid in 1:1:1 ratio) was added to the reaction mixture and incubated at 25°C for 150 minutes.

1.6.1.3 Pink colored complex formed was read at 540nm.

1.6.1.4 The nitric oxide scavenging activity of antioxidant/radioprotector compound was calculated as percent decreased in the absorbance of the complex formed by diazotization of nitrite with sulfanilamide and subsequent coupling with naphthylethylenediamine. Decrease in absorbance (OD) was an indicative of higher nitric oxide radical scavenging potential of a test compound.

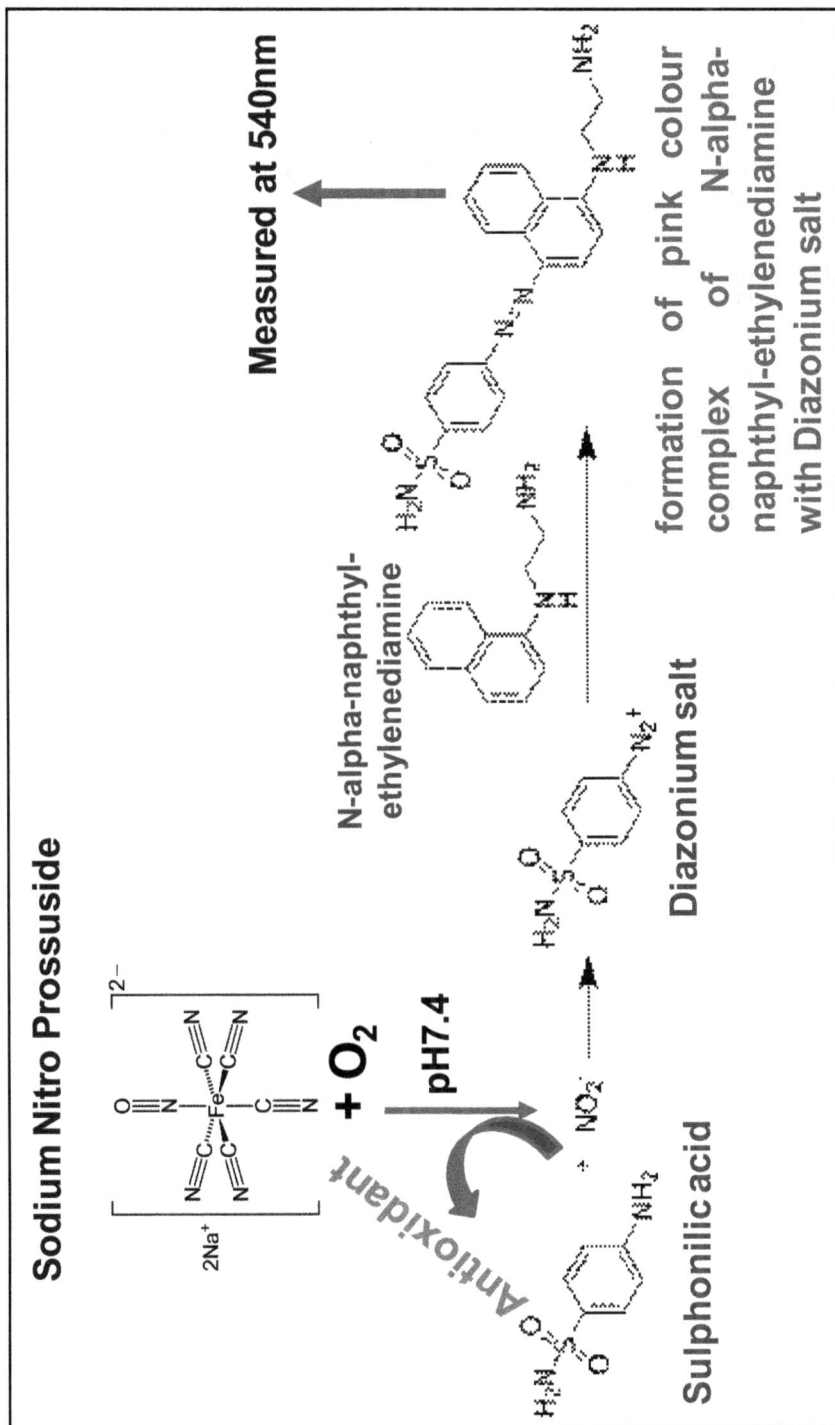

Figure 6: Principle and Chemical Mechanism of Nitric Oxide Radicals Scavenging by Antioxidant *In vitro* System.

References

1. Banerjeen S, Chanda A, Ghoshal A, Debnath R, Chakraborty S, Saha R and Das A. (2011). *Ixora Coccinea* Nitric Oxide Scavenging Activity Study of Ethanolic Extracts from Two Different Areas of Kolkata. *Asian J. Exp. Biol. Sci.* 4, 595-599.

2. Singh D, Mishra M, Gupta M, Singh P, Gupta A and Nema R (2012). Nitric Oxide radical scavenging assay of bioactive compounds present in methanol Extract of *Centella asiatica. Intern. J. Pharm. Pharmaceut. Scie. Res.* 2(3), 42-44.

3. Parul R, Kundu S K and Saha P (2012). *In vitro* Nitric Oxide Scavenging Activity of Methanol Extracts of Three Bangladeshi Medicinal Plants. *Pharma Innovation.* 1(12), 83-88.

4. Pooja, Sharma P, Samanta KC, Garg V (2010). Evaluation of nitric oxide and hydrogen peroxide scavenging activity *Dalbergia sissoo* roots. *Pharmacophore.* 1(2), 77-81.

5. Chakraborthy G S (2009). Free radical scavenging activity of *Aesculus indica* leaves. 1(3), 524-526.

6. Olabinri B, Odedire OO, Olaleye MT, Adekunle A S, Ehigie L O and Olabinri P F (2010). *In vitro* evaluation of hydroxyl and nitric oxide radical scavenging activities of artemether. 5(1), 102-105.

Protocol No. 1.7: Colorimetric Estimation of Superoxide Radicals Scavenging Properties of Radioprotective Compounds

Assay Principle

Dimethylsulfoxide (DMSO) in alkaline condition generates superoxide radicals. These superoxide radicals if not scavenged/neutralized by antioxidant reagent, further oxidize the nitrobluetetrazoliun (NBT). Followed by reduction, NBT form a formazen complex of purple colour (Figure 7) that can be measured at 560nm using spectrophotometer (1-6). Antioxidant able to neutralize superoxide are able to reduce formazen formation in vitro.

Assay Requirement

Sodium hydroxide, nitroblue tetrazolium (NBT), dimethylsulphoxide (DMSO), spectrophotometer, eppondrof tube (2.0 ml), micro-pipette

1.7.1 Assay Procedure

1.7.1.1 Superoxide radicals were generated from air saturated dimethylsulphoxide (DMSO; 1 ml) by addition of sodium hydroxide (1.0mM; 0.5ml) to DMSO.

1.7.1.2 Superoxide radicals may remain stable in solution, which reduces nitrobluetetrazolium (NBT) into formazan dye at room temperature that can be measured at 560 nm.

1.7.1.3 Briefly, the reaction mixture containing 1.0 ml of alkaline DMSO (1.0 ml DMSO containing 5 mM NaOH in 0.1 ml water) and 0.3 ml of the antioxidant agents or compound/extract under evaluation in freshly distilled DMSO at various concentrations.

1.7.1.4 0.1 ml of NBT (1.0 mg/ml) was added to the reaction mixture to give a final volume of 1.4 ml.

1.7.1.5 The reaction mixture incubated at room temperature for 5-10 minutes and the purple colour appeared was measured at 560 nm.

References

1.   Noda Y, Anzai K, Marl A, Kohno M, Shinmei M and Packer L (1997). Hydroxyl and superoxide anion radical scavenging activities of natural source antioxidants using the computerized jes-fr30 ESR spectrometer system. *Biochem. Mol. Biol. Intern.* 42, (1), 35-44.

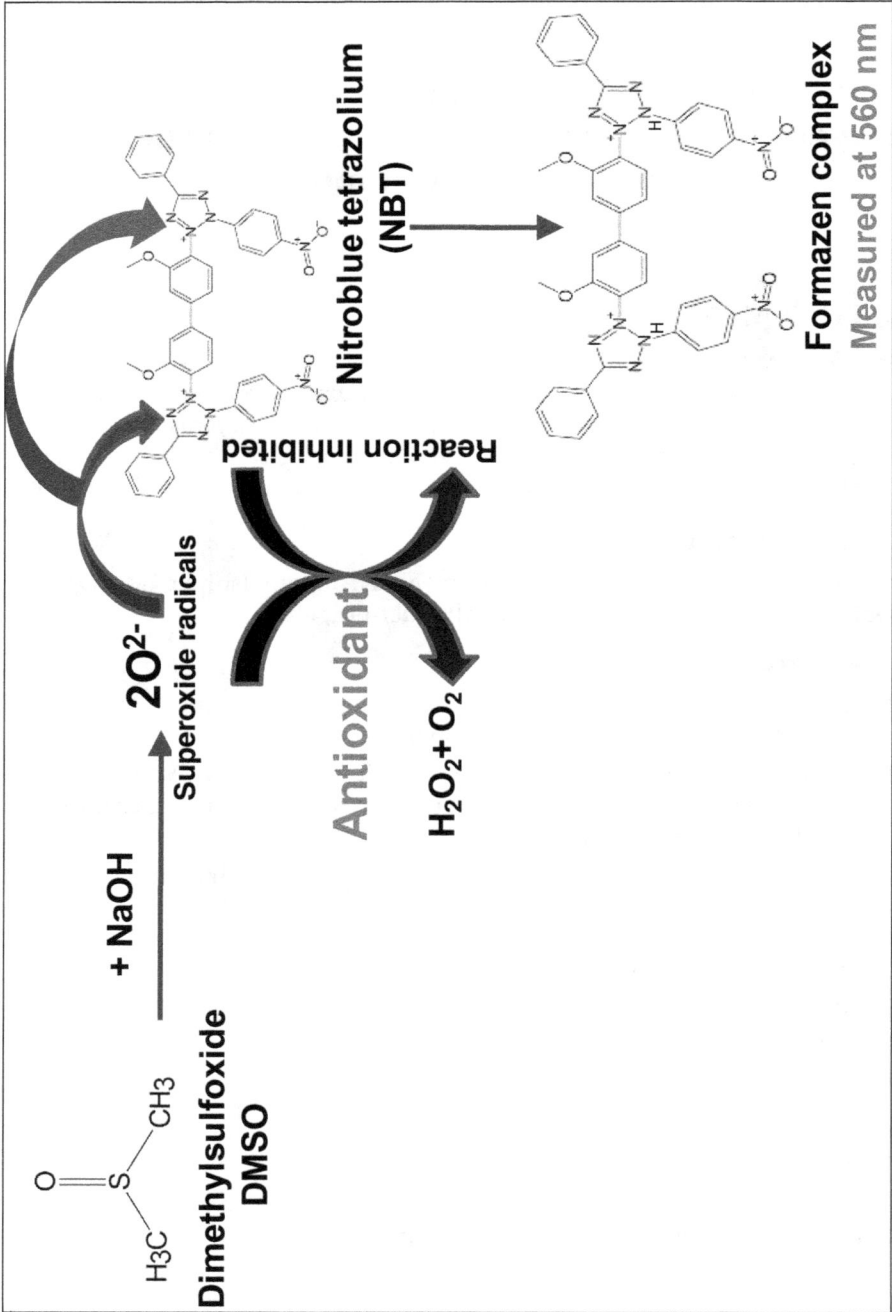

Figure 7: Reaction Mechanism Depicted Superoxide Radicals Generation, Formazen Complex Formation and Superoxide Radicals Scavenging by Antioxidant Agent.

2. Nandi A and Chatterjee I B (1987). Scavenging of superoxide radical by ascorbic acid. *J. Biosci.* 11, 435–441.

3. Tao H, Zhou J, Wu T, and Cheng Z (2014). High-throughput superoxide anion radical scavenging capacity assay. *J. Agric. Food Chem.* 62(38), 9266–9272.

4. Sánchez-Moreno C (2002). Review: Methods used to evaluate the free radical scavenging activity in foods and biological systems. *Food Sci. Technol. International.* 8 (3), **121-137.**

5. Wang I, Liu S, Zheng Z, Pi Z, Song F and Liu Z (2014). Rapid assay for testing superoxide anion radical scavenging activities to natural pigments by ultra-high performance liquid chromatography-diode-array detection method. DOI:10.1039/C4AY02690J.

6. Suzumura K, Yasuhara M and Narita H (1999). Superoxide anion scavenging properties of fluvastatin and its metabolites. *Chem. Pharm. Bull.* 47(10), 1477-1480.

Protocol No. 1.8: Colorimetric Estimation of Copper (II) Reduction Capacity (CUPRAC) of Radioprotective Compounds/ Biological Fluid/Plant Extract etc.

Principle of the Assay

The CUPRAC (Cupric Reducing Antioxidant Capacity) assay of antioxidant measurement, is based on the absorbance measurement of Cu(I)-neocuproine (Nc) chelate complex formation. Cu^{+2} radicals generated from the $CuCl_2$ can reduce Cu^{+2} to Cu^{+1}. Reduced Cu^{+1} can easily be chelated with neocuproine and form a chromogenic complex measured by spectrophotometrically at 450 nm (1-7). Colour intensity of the chromogenic complex (Figure 8) is directly proportional to the total antioxidant activity of the test compound.

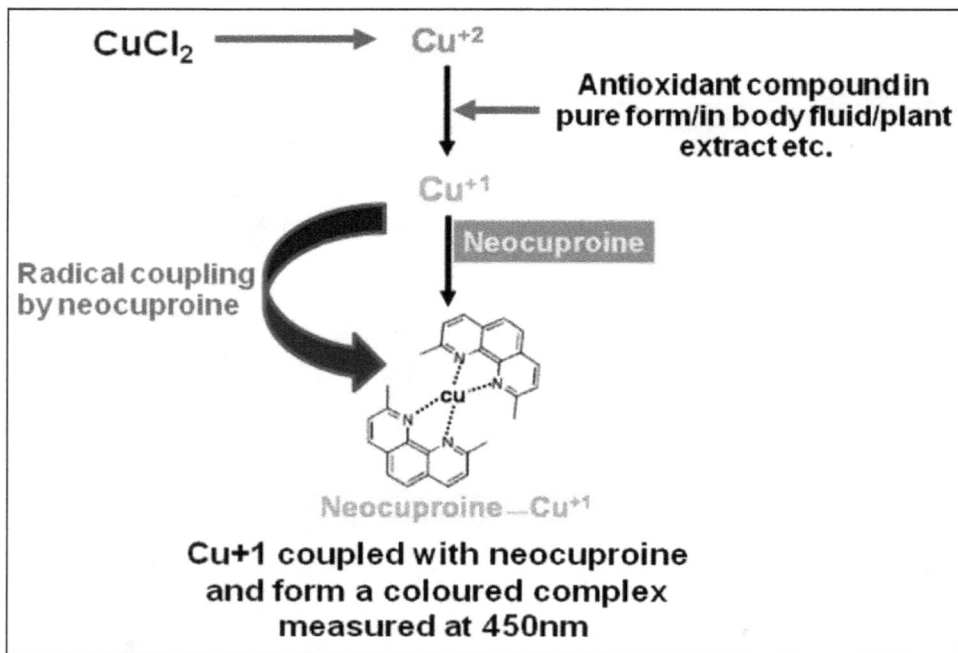

Figure 8: Diagrametical Representation of Chemical Mechanistic of the CUPRAC Assay.

Assay Requirement

CuCl$_2$, neocuproine, alcohol, ammonium acetate (NH$_4$Ac) buffer, 5ml centrifuge tube, centrifuge, UV-Vis spectrophotometer

1.8.1 Assay Procedure

1.8.1.1 Briefly, Reaction mixture was prepared by mixing 1 ml of working solution of 0.02M CuCl$_2$; 1ml of 0.0075 M neocuproine alcoholic solution and 1ml of ammonium acetate (NH$_4$Ac) buffer solution (1:1:1 ratio.).

1.8.1.2 Reaction component were mixed with 12.5 µl of test sample in a centrifuge tube.

1.8.1.3 All reaction tubes were centrifuged for 3.0 minutes at 4000 rpm.

1.8.1.4 Supernatant was collected and incubated for 30 minutes at room temperature for absorbance measurements.

1.8.1.5 The absorbance of the colored complex formed was measured against the blank containing all components without test compound at 450nm.

References

1. Apak R, Güçlü K, Ozyürek M, Karademir SE and Altun M (2005).Total antioxidant capacity assay of human serum using copper(II)-neocuproine as chromogenic oxidant: the CUPRAC method. *Free Radic Res.* 39(9): 949-961.

2. Apak R, Güçlü K, Ozyürek M, Bektaþoðlu B and Bener M (2010). Cupric ion reducing antioxidant capacity assay for antioxidants in human serum and for hydroxyl radical scavengers.

3. Apak R, Güçlü K, Ozyürek M and Karademir SE. (2004). Novel total antioxidant capacity index for dietary polyphenols and vitamins C and E, using their cupric ion reducing capability in the presence of neocuproine: CUPRAC method. *J Agric Food Chem.* 52(26), 7970-7981.

4. Apak R, Güçlü K, Ozyürek M, Bektas Oðlu B and Bener M (2008).Cupric ion reducing antioxidant capacity assay for food antioxidants: vitamins, polyphenolics, and flavonoids in food extracts.*Methods Mol Biol.* 477,163-193.

5. Apak R, Güçlü K, Ozyürek M, Esin Karademir S and Erçað E. (2006). The cupric ion reducing antioxidant capacity and polyphenolic content of some herbal teas. *Int J Food Sci Nutr.* 57, 292-304.

6. Ozyürek M, Bektaþoðlu B, Güçlü K, Güngör N and Apak R(2008). Simultaneous total antioxidant capacity assay of lipophilic and hydrophilic antioxidants in the same acetone-water solution containing 2 per cent methyl-beta-cyclodextrin using the cupric reducing antioxidant capacity (CUPRAC) method. *Anal Chim Acta.* 7, 630(1), 28-39

7. Send Cekiç SD, Bakan KS, Tütem E and Apak R (2009). Modified cupric reducing antioxidant capacity (CUPRAC) assay for measuring the antioxidant capacities of thiol-containing proteins in admixture with polyphenols. *Talanta.* 79(2), 344-351.

Protocol No. 1.9: Spectrophotometric Estimation of N,N-dimethyl-p-Phenylenediamine (DMPD) Radicals Scavenging by a Radioprotective/ Antioxidant Compound

Assay Principle

N,N-dimethyl-p-Phenylenediamine (DMPD) reacts with potassium persulfate and generate a $DMPO^{2+}$ radicals that can be measured directly at 517 nm spectrophotometrically. However, if antioxidant molecule able to reduced $DMPO^{+2}$ radicals in the reaction environment, the concentration of the $DMPO^{+2}$ radicals will be reduced lead to significant reduction in absorbance at 517 nm (Figure 9). Therefore, antioxidant activity of the test compound will be inversely propotional to the decreasing absorbance of $DMPO^{+2}$ solution in presence of antioxidant in the reaction.

Assay Requirement

N,N-dimethyl-p-Phenylenediamine (DMPD), deionized water, potassium persulfate, acetate buffer (pH5.6), Uv-Vis spectrophotometer.

1.9.1 Assay Procedure

1.9.1.1 100mM solution of N,N-dimethyl-p-Phenylenediamine (DMPD) was prepared by solubilizing 0.209 gm of DMPO in 10 ml of deionized water.

1.9.1.2 50 µl of 0.4 mM potassium persulfate solution was mixed with 100 µl of DMPO solution to generate the DMPO radicals *in vitro* system. Final volume of the reaction mixture was made 10 ml with the help of acetate buffer (pH 5.6).

1.9.1.3 The reaction mixture was incubated at 25 °C for 3hr in the dark before using as free radical source and the DMPO radicals formed during reaction stored at 4 °C in the dark for several hours.

1.9.1.4 225 µl of DMPO was transferred directly to the microwell and absorbance at 517 nm was measured (A0).

1.9.1.5 15µl test samples were added to all wells.

1.9.1.6 210 µl of DMPD+ radical solution was added to all samples wells and plates were stirred and left to stand for 10 minutes.

1.9.1.7 A decrease in absorbance was measured (A1).

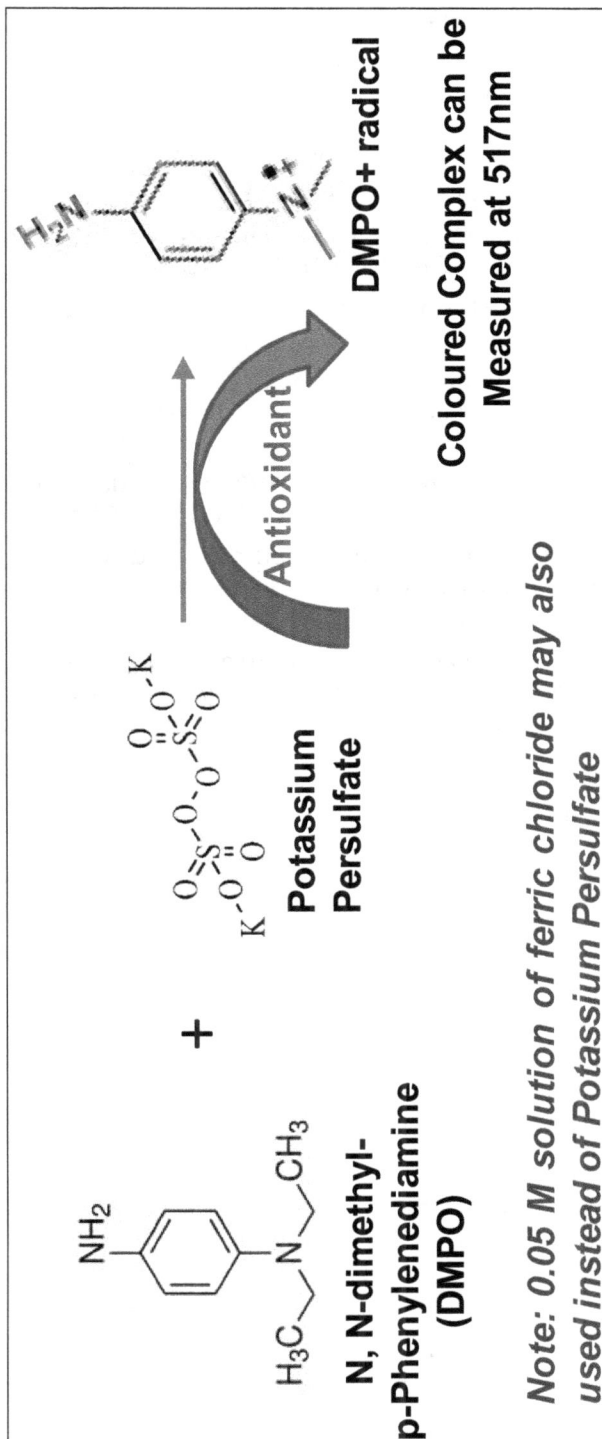

Figure 9: Chemical Principle of DMPO^{+2} Radicals Generation and their Neutralization by Antioxidant Compound.

1.9.1.8 The pure DMPD.solution was used as control [A0].

1.9.1.9 The DMPD⁺ radicals scavenging effect of the test compound was calculated using following equation:= [A0 −A1 / A0]×100

References

1. Asghar MN, Khan IU, Sherin L, Muhammad N A and Sherin L (2007). Evaluation of antioxidant activity using an improved DMPD radical cation decolorization assay. *Acta Chim. Slov.* 54, 295–300.

2. Ak T and Gülçin I (2008). Antioxidant and radical scavenging properties of curcumin. *Chem. Biol. Interact.* 174(1), 27-37.

3. Apak R, Gorinstein S, Böhm V, Schaich K M, Özyürek M and Güçlü K (2013). Methods of measurement and evaluation of natural antioxidant capacity/ activity (IUPAC Technical Report). *Pure Appl. Chem.* 85(5), 957–998.

4. Vijaya Kumar A, Kumar PP and Jeyaraj B (2013). Antioxidant activity of *Illicium griffithi* hook. F. and Thoms seeds - *in vitro*. *Asian J Pharm Clin Res*, 6(2), 269-273.

5. Sánchez-Moreno C (2002). Review: Methods used to evaluate the free radical scavenging activity in foods and biological systems. *Food Sci. Technol. International*, 8: 121-137.

6. Kalava V S, and Menon SG (2012). *In vitro* free radical scavenging activity of aqueous extract from the mycelia of V*olvariella volvacea* (bulliard ex fries) singer. *Int J. Curr. Pharm. Res.* 4(3), 94-100.

Section 2

Secondary Methods to Screen Out Radioprotective Compounds

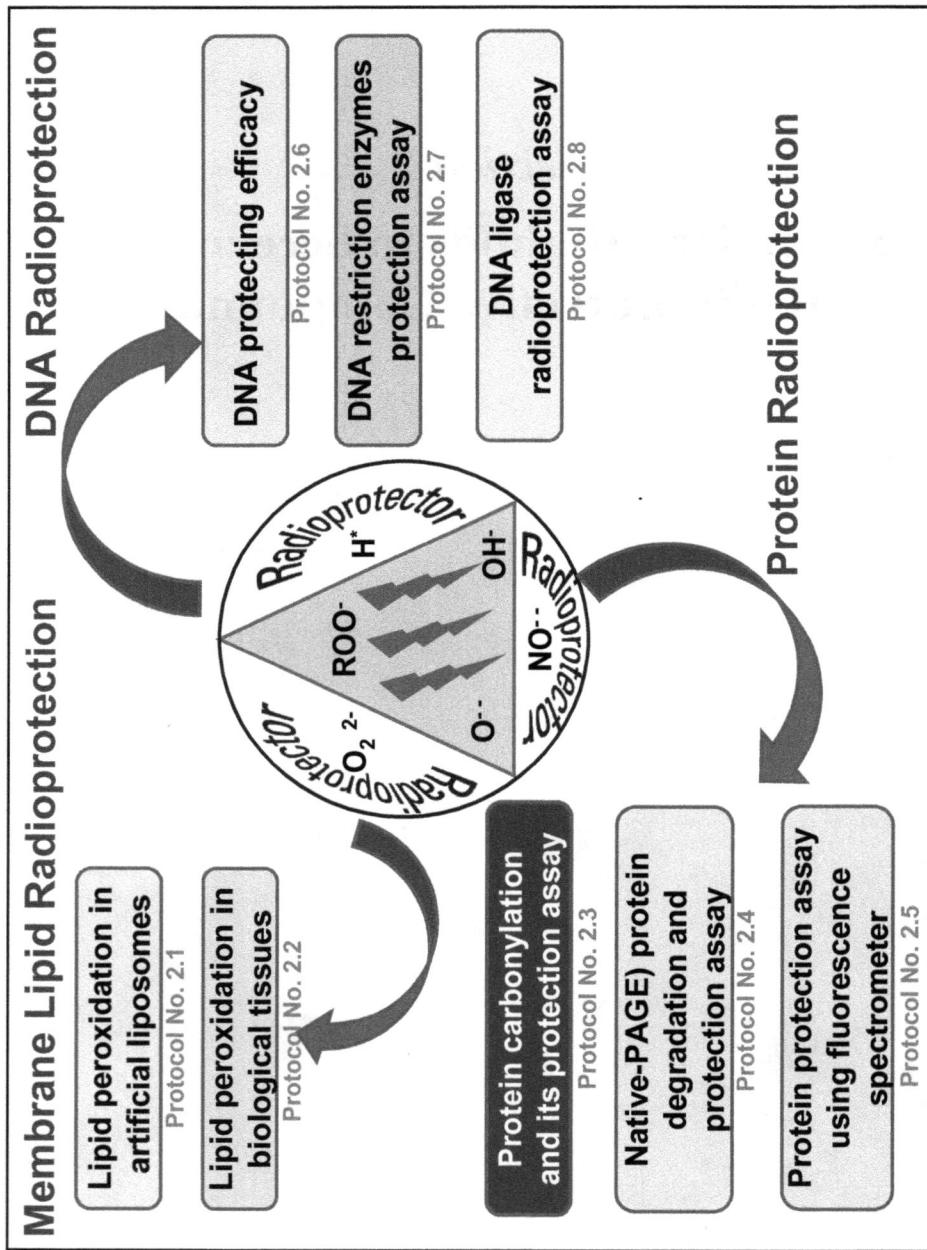

Figure 10: Schematic Summary of Secondary Methods of Radioprotective Drug Screening.

Protocol No. 2.1: Screening of Radioprotective Compounds able to Inhibit Radiation Induced Lipid Peroxidation in Artificial Liposomes

2.1.1 Artificial Liposome Preparation

2.1.1.1 A solution of soya lecithin and cholesterol was mixed in 1:3 ratio with 5.0 ml of chloroform in a round bottom flask.

2.1.1.2 The solvent was removed under pressure of 200psi using rotatory vacuum evaporator at 37°C with continuous rotation at 60rpm.

2.1.1.3 Thin lipid film formed at the wall of the round bottom flask was dispersed in the 0.1M PBS with continuous shaking at 40°C.

2.1.1.4 The resultant milky suspension containing multi-lamellar liposomes was aliquoted and stored at 4°C for future experiments.

2.1.2 Anti-lipid Peroxidation Assay

2.1.2.1 Different concentration (25-2000µg/ml) of test compound was added with 100µl of liposome suspension and incubated for 30 minutes at 37°C.

2.1.2.2 A control set of reaction mixture containing liposomes (100µl) without test compound (replaced with 1.0 ml PBS) was also prepared and incubated similarly.

2.1.2.3 Both test and control reaction mixtures were exposed to the gamma radiation (not less than 250 Gy).

2.1.2.4 Irradiated samples were than re-incubated at 37°C for 1h.

2.1.2.5 Followed by incubation, 600µl of 0.1M PBS (pH 7.4) and 100µl of 10 per cent TCA was added to the reaction mixture.

2.1.2.6 Samples were centrifuged at 1000Xg for 10 minutes at room temperature and supernatant separated out.

2.1.2.7 100µl of TBA (0.67 per cent) was added to the supernatant and samples kept for boiling water bath till the pink coloured-MDA-complex formed and stabilized.

2.1.2.8 Absorbance of the reaction mixture was recorded at 535nm and compared with the controls. Percent decrease in the absorbance with the irradiated liposomes pretreated with antioxidant compound as compared to irradiated control (without drug) was directly proportional to the percentage of lipid peroxidation inhibition.

References

1. Gupta V, Misra V and Viswanathan N P (1989). Artificial liposomes as a model for studying the lipid peroxidation effect of p-phenylene diamine. *J Microencapsul.* 6(3), 311-318.

2. Kaurinovic B and Popovic M (2012). Liposomes as a tool to study lipid peroxidation. Kaurinovic and Popovic, licensee In: Tech publication. pp-155-180. http://dx.doi.org/10.5772/46020.

3. Cholbi MR, Paya M and Alcaraz MJ (1991). Inhibitory effect of phenolic compounds on CCl_4-induced microsomal lipid peroxidation. *Experimentia.* 47, 195-198.

4. Rekka E and Kourounakis PN (1991). Effect of hydroxyethyl rutoside and related compounds on lipid peroxidation and free radical scavening activity: Some structural aspects. *J. Pharmac. Pharmacol.* 43, 486-490.

5. Doba T, Burton GW and Ingold KU (1985). Antioxidant and co-oxidant activity of Vitamin C: The effects of vitamin C, either alone or in the presence of vitamin E or a water-soluble vitamin E analogue, upon the peroxidation of aqueous multimalleral phospholipid liposomes. *Biochim. Biophys. Acta.* 835, 298-303.

6. Immordino M L, Dosio F and Cattel L (2006). Stealth liposomes: Review of the basic science, rationale, and clinical applications, existing and potential. *IJN* 1(3): 297-315.

7. Chatterjee S N and Agarwal S (1988). Liposomes as membrane model for study of lipid peroxidation. *Free Rad. Biol. Med.* 4, 51-72.

8. Cighetti G, Allevi P, Debiasi S and Paroni R (1997). Inhibition of *in vitro* lipid peroxidation by stable steroidic nitroxyl radicals. *Chem. Phy. Lipids* 88, 97–106.

Protocol No. 2.2: Screening of Radioprotective Compounds able to inhibit Radiation Induced Lipid Peroxidation in Biological Tissues

Assay Requirement

Phosphate buffer saline (pH 7.4), trichloroacetic acid, thiobarbituric acid (TBA), micro-centrifuge tubes, gamma irradiator and spectrophotometer.

2.2.1 Preparation of Brain and Liver Homogenate

2.2.1.1 Whole brain and liver portion of the $C_{57}BL/6$ or Swiss albino mice was excised, collected and stored in the phosphate buffer saline (0.1M).

2.2.1.2 10 mg of brain or liver tissue sample was mixed with 10 ml of PBS (pH 7.0).

2.2.1.3 Tissue homogenate was prepared with the help of hand held homogenizer.

2.2.1.4 The tissue homogenate was centrifuged of 5000rpm at 4°C temperature and supernatant separated for experimentation.

2.2.2 Anti-lipid Peroxidation Assay

2.2.2.1 Different concentration ((25-2000µg/ml) of test compound was added with 100 µl of 10 per cent of brain/liver tissue homogenate and incubated for 30 minutes at 37°C.

2.2.2.2 A control sample of tissue homogenate without test compound will also be prepared and incubated similarly.

2.2.2.3 Both test and control reaction mixtures were exposed to the γ radiation (not less than 250Gy).

2.2.2.4 Irradiated samples were-incubated at 37°C for 1h.

2.2.2.5 Followed by incubation, 600µl of 0.1M PBS (pH 7.4) and 100µl of 10 per cent TCA was added to the reaction mixture.

2.2.2.6 Samples were centrifuged at 1000Xg for 10 minutes at room temperature and supernatant separated out.

2.2.2.7 100µl of TBA (0.67 per cent) was added to the supernatant and samples kept for boiling in water bath until pink coloured- MDA- complex formed and stabilized.

2.2.2.8 Absorbance of samples was recorded at 535nm and compared with the controls. Percent decrease in the absorbance with the irradiated tissue homogenate pretreated with antioxidant compound as compared to irradiated control (without drug treatment) was directly proportional to the percentage of lipid peroxidation inhibition (1-8).

References

1. Gupta V, Misra V and Viswanathan PN (1989). Artificial liposomes as a model for studying the lipid peroxidation effect of p-phenylene diamine. *J Microencapsul.* 6(3), 311-318.

2. Kaurinovic B and Popovic M (2012). Liposomes as a tool to study lipid peroxidation. Kaurinovic and Popovic, licensee In: Tech publication. pp-155-180. http://dx.doi.org/10.5772/46020.

3. Cholbi MR, Paya M and Alcaraz MJ (1991). Inhibitory effect of phenolic compounds on CCl_4-induced microsomal lipid peroxidation. *Experimentia.* 47, 195-198.

4. Rekka E, Kourounakis PN (1991). Effect of hydroxyethyl rutoside and related compounds on lipid peroxidation and free radical scavening activity: Some structural aspects. *J. Pharmac. Pharmacol.* 43, 486-490.

5. Doba T, Burton GW and Ingold KU (1985). Antioxidant and co-oxidant activity of Vitamin C: The effects of vitamin C, either alone or in the presence of vitamin E or a water-soluble vitamin E analogue, upon the peroxidation of aqueous multimalleral phospholipid liposomes. *Biochim. Biophys. Acta.* 835, 298-303.

6. Immordino M L, Dosio F and Cattel L (2006). Stealth liposomes: Review of the basic science, rationale, and clinical applications, existing and potential. *IJN* 1(3): 297-315.

7. Chatterjee S N and Agarwal S (1988). Liposomes as membrane model for study of lipid peroxidation. *Free Rad. Biol. Med.* 4, 51-72.

8. Cighetti G, Allevi P, Debiasi S and Paroni R (1997). Inhibition of *in vitro* lipid peroxidation by stable steroidic nitroxyl radicals. *Chem. Phy. Lipids* 88, 97–106.

Protocol No. 2.3: Screening of Compounds Protect against Gamma Radiation Induced *In vitro* Protein Carbonylation

Assay Requirement

Bovine serum albumin, phosphate buffer saline (pH 7.0), DNPH, hydrochloric acid, trichloric acid, guanidine-HCl, microcentrifuge tube (2.0 ml), centrifuged and spectrophometer.

2.3.1 Protein Protection Assay

2.3.1.1 Bovine Serum Albumin (BSA) stock solution (1.0 mg/ml) was prepared.

2.3.1.2 BSA stock solution was divided into two different sets *i.e.* Test (set 1) and control sample (Set 2).

2.3.1.2 In set 1, 250µl of BSA stock solution (1.0 mg/ml) was mixed with equal amount and volume of radioprotective compound.

2.3.1.3 In set 2, BSA samples added without radioprotective compounds and final volume (1.0 ml) was adjusted equally with PBS.

2.3.1.4 Now, both the sets were irradiated with different doses (400-3000Gy) of gamma radiation.

2.3.1.5 250µl of 10mM DNPH in 2.5M HCl was added in set 1 (test) samples. While DNPH was replaced with 2.5 M HCl in set 2 (control) samples.

2.3.1.6 All sets were incubated in the dark for 15 minutes along with vortexing at every 5 minutes.

2.3.1.7 125µl of 50 per cent (w/v) TCA was added to both the sets of samples and then incubated at 20°C for at least 15 minutes.

2.3.1.8 After incubation, all the samples were centrifuged at 9000xg for 15 minutes at 4°C. Supernatant was discarded without disturbing the pellet.

2.3.1.9 The pellet was than washed thrice with ice cold ethanol-ethyl-acetate (1:1), centrifuged for 2 minutes at 9000xg and supernatant discarded after each wash.

2.3.1.10 Followed by washing, pellet was dissolved in 6M guanidine-HCl and absorbance read at 370nm taking 6M guanidine-HCl as blank (1-13).

2.3.1.11 Total carbonyl groups generated in irradiated samples in presence or absence of radioprotective compound was calculated in terms of nM carbonyl/mg protein.

References

1. Dalle-Donne I, Rossi R, Giustarini D, Milzani A and Colombo R (2003). Protein carbonyl groups as biomarkers of oxidative stress. *Clinica. Chimica. Acta.* 329, 23–38.

2. Dean RT, Fu S, Stocker R and Davies MJ (1997). Biochemistry and pathology of radical-mediated protein oxidation. *Biochem. J.* 324, 1 –18.

3. Berlett BS and Stadtman ER (1997). Protein oxidation in aging, disease, and oxidative stress. *J. Biol. Chem.* 272, 20313– 20316.

4. Stadtman ER and Berlett BS (1997). Reactive oxygen-mediated protein oxidation in aging and disease. *Chem. Res. Toxicol.*10, 485– 494.

5. Chevion M, Berenshtein E and Stadtman ER (2000). Human studies related to protein oxidation: protein carbonyl content as a marker of damage. *Free. Radic. Res.* 33, S99– S108.

6. Levine RL, Garland D, Oliver CN, Amici A, Climent I and Lenz A (1990). Determination of carbonyl content in oxidatively modified proteins. *Methods Enzymol.* 186, 464–478.

7. Lenz A, Costabel U, Shaltiel S and Levine L (1989). Determination of carbonyl groups in oxidatively modified proteins by reduction with tritiated sodium borohydride. *Anal. Biochem.* 177, 419– 425.

8. Yan LJ and Sohal R S (1998). Gel electrophoretic quantitation of protein carbonyls derivatized with tritiated sodium borohydride. *Anal. Biochem.* 265, 176– 182.

9. Levine RL, Williams J, Stadtman ER and Shacter E (1994). Carbonyl assays for determination of oxidatively modified proteins. *Methods Enzymol.* 233, 346– 357.

10. Fagan JM, Sleczka BG and Sohar I (1999). Quantitation of oxidative damage to tissue proteins. *Int. J. Biochem. Cell Biol.* 31, 751–757.

11. Dalle-Donne I, Rossi R, Giustarini D, Gagliano N, Lusini L, Milzani A, *et al.* (2001). Actin carbonylation: from a simple marker of protein oxidation to relevant signs of severe functional impairment. *Free Radic. Biol. Med.* 31, 1075–1083.

12. Reznick AZ and Packer L (1994). Oxidative damage to proteins: spectrophotometric method for carbonyl assay. *Methods Enzymol.* 233, 263–357.

13. Winterbourn CC, Buss IH, Chan TP, Plank LD, Clark MA and Windsor JA (2000). Protein carbonyl measurements show evidence of early oxidative stress in critically ill patients. *Crit. Care Med.* 28(1), 143–149.

Protocol No. 2.4: Evaluation of Radioprotective Efficacy of Test Compound against Gamma Radiation Induced Damage to Protein (BSA): Native Poly-Acrylamide Gel Electrophoresis (PAGE) Analysis

Assay Requirement

Bovine serum albumin, Milli Q water, acrylamide, bis-acrylamide, Tris-HCl, glycerol, bromophenol blue, Coomassie brilliant blue R-250, methanol, glacial acetic acid, electrophoresis unit along with power supply and gamma irradiator.

2.4.1 Assay Procedure

2.4.1.1 Stock solution (1.0 mg/0.1ml) of BSA was prepared.

2.4.1.2 To analyze the structural damage induced by gamma radiation, three different experimental sets were prepared:

Set 1: 10µl of BSA solution in milli Q water (control).

Set 2: 10µl of BSA solution in milli Q water irradiated with 600, 800, 1000 and 1200 Gy of gamma radiation individually.

Set 3: 10µl of BSA solution in milli Q water irradiated in presence of radioprotective compound in 2:1 ratio (*i.e.*10µl BSA+ 5µl test compound + irradiation; 600, 800, 1000 and 1200Gy in individual subsets).

2.4.1.3 Denaturing 10 per cent polyacrylamide gel was prepared simultaneously.

2.4.1.4 5.0µl of sample buffer (0.5M Tris-HCl, pH6.8; 5 per cent v/v glycerol; 0.5 per cent v/v bromophenol blue) was mixed with all sets of the irradiated and control samples.

2.4.1.5 Samples were then loaded on SDS-PAGE gel.

2.4.1.6 Electrophoresis was carried out at a constant voltage (stacking at 60V and resolving at 100V).

2.4.1.7 After electrophoresis, the gels was stained with 0.1 per cent Coomassie brilliant blue R-250 in methanol: glacial acetic acid: water (4:2:4) with gentle shaking at room temperature for at least 5h.

2.4.1.8 Followed by destaining with the washing solution (methanol: acetic acid: water, 1.0:0.7:8.3) distinct bands over a clear background was observed.

2.4.1.9 Increasing smearing of BSA bands with increasing radiation doses in absence of radioprotective compound was reflected as radiation induced structural damage to BSA.

2.4.1.10 Whereas, decreased smearing in the protein bands of irradiated BSA pretreated with radioprotective compound corresponded to the protection offered by radioprotective compound against gamma radiation.

References

1. Yan L J and Sohal R S (1998). Gel electrophoretic quantitation of protein carbonyls derivatized with tritiated sodium borohydride. *Anal. Biochem.* 265,176– 182.

2. Shacter E (2000). Protein oxidative damage. *Methods Enzymol.* 319, 428– 436.

3. Aksenov M Y, Aksenova M V, Butterfield D A, Geddes J W and Markesbery W R (2001). Protein oxidation in the brain in Alzheimer's disease. *Neuroscience.* 103, 373– 383.

4. Fu S, Davies M J, Stocker R and Dean R T (1998). Evidence for roles of radicals in protein oxidation in advanced human atherosclerotic plaque. *Biochem. J.* 333, 519– 525.

5. Griffiths H R (2000). Antioxidant and protein oxidation. *Free Radic. Res.* 33, S47–S58.

6. Lim P S, Cheng Y M and Wei Y H (2002). Increase in oxidative damage to lipids and proteins in skeletal muscle of uremic patients. *Free Radic. Res.* 36(3), 295-301.

7. Chen S S, Chang L S and Wei Y H (2001). Oxidative damage to proteins and decrease of antioxidant capacity in patients with varicocele. *Free Radic. Biol. Med.* 30(11), 1328–1334.

Protocol No. 2.5: Evaluation of Radioprotective Efficacy of a Test Compound against Gamma Radiation Induced Damage to Protein (BSA) using Fluorescence Spectrophotometric Analysis

Assay Requirement

Bovine serum albumin, Milli-Q water microtubes, fluorescence spectrophotometer and gamma irradiator.

2.5.1 Assay Procedure

2.5.1.1 Stock solutions (1mg/ml) of Bovine serum albumin (BSA) and radioprotective compound were prepared.

2.5.1.2 Three different sets of samples were prepared:

Set 1: BSA samples were irradiated with different doses (600Gy, 800Gy, 1000Gy and 1200Gy individually) of gamma radiation.

Set 2: Test compound (20µl) was mixed with BSA (980µl) and mixture was irradiated with different doses of gamma radiation (*i.e.* 600Gy, 800Gy, 1000Gy and 1200Gy individually).

Set 3: Control unirradiated BSA

2.5.1.3 Followed by irradiation, all samples were analyzed fluorospectrometrically using excitation 280nm and emission wavelength between 300-500 nm.

2.5.1.4 A constant increase in BSA emission spectra along with increasing doses of gamma radiation in absence of radioprotective compound will reflect the radiation induced damage to BSA.

2.5.1.5 While reduction in emission spectral intensity of the irradiated BSA in presence of radioprotective compound,suggested protection offered by the radioprotective compound to the BSA against radiation induced damage (1-7).

References

1. Yan L J and Sohal R S (1998). Gel electrophoretic quantitation of protein carbonyls derivatized with tritiated sodium borohydride. *Anal. Biochem.* 265, 176–182.

2. Shacter E (2000). Protein oxidative damage. *Methods Enzymol.* 319, 428– 436.

3. Aksenov M Y, Aksenova M V, Butterfield D A, Geddes J W and Markesbery W R (2001). Protein oxidation in the brain in Alzheimer's disease. *Neuroscience.* 103, 373– 383.

4. Fu S, Davies M J, Stocker R and Dean R T (1998). Evidence for roles of radicals in protein oxidation in advanced human atherosclerotic plaque. *Biochem. J.* 333, 519– 525.

5. Griffiths H R (2000). Antioxidant and protein oxidation. *Free Radic. Res.* 33, S47–S58.

6. Lim P S, Cheng Y M and Wei Y H (2002). Increase in oxidative damage to lipids and proteins in skeletal muscle of uremic patients. *Free Radic. Res.* 36(3), 295-301.

7. Chen S S, Chang L S and Wei Y H (2001). Oxidative damage to proteins and decrease of antioxidant capacity in patients with varicocele. *Free Radic. Biol. Med.* 30(11),1328–1334.

Protocol No. 2.6: Evaluation of Plasmid DNA Protecting Efficacy of Radioprotective Compound against Gamma Radiation Induced Damage

Assay Requirement

Plasmid DNA (pUC19), Tris-EDTA buffer, agarose, ethidium bromide, DNA electrophoresis unit, UV trans-illuminator and densitometric analysis software.

2.6.1 Assay Procedure

2.6.1.1 Plasmid DNA pUC19 (250ng) was mixed with varied concentrations (2.5-5000 μg) of test compound in Tris-EDTA (pH 8.0, 30mM) buffer individually.

2.6.1.2 A separate set of plasmid DNA (250ng) without test compound was prepared simultaneously.

2.6.1.3 A fix dose of 250Gy of gamma irradiation was delivered to both the sets of reaction mixtures using gamma irradiator and incubated for 1h at 37°C.

2.6.1.4 Followed by incubation, plasmid DNA was mixed with loading dye.

2.6.1.5 Reaction mixture was loaded onto 1 per cent agarose gel and electrophoresis carried out at 60V using DNA submarine electrophoresis unit.

2.6.1.6 Upon completion of electrophoresis, gel was stained with ethidium bromide (0.5μg/ml) for 30 minutes.

2.6.1.7 DNA bands were visualized under UV trans-illuminator.

2.6.1.8 Densitometric analysis was performed using densitometric analysis software.

2.6.1.9 The percentage of super-coiled form (per cent SC) of plasmid DNA (front running bands) was represented the protection against single-strand breaks.

2.6.1.10 Whereas per cent open circular form (per cent OC) of plasmid DNA (Back running bands) was represented DNA damage (1-8).

2.6.1.11 The percentage of DNA protection was calculated using integrated density value (IDV) formula as:

Per cent SC form = $(SC_{IDV}/SC_{IDV} + OC_{IDV})$ x 100 (per cent)

References

1. Alok A, Adhikari J S and Chaudhury N K (2013). Radioprotective role of clinical drug diclofenac sodium. *Mutation Res.* 755 156–162.

2. Maurya DK, Salvi VP and Nair C K K (2005). Radiation protection of DNA by ferulic acid under *in vitro* and *in vivo* conditions. *Mol. Cell. Biochem.* 280, 209-217.

3. Denison L, Haigh A, D'Cunha G and Martin R F (1991). Erratum: DNA ligands as radioprotectors: Molecular studies with Hoechst 33342 and Hoechst 33258. Intn. J. Rad. Biol. 61, 69-81.

4. Newton G L, Ly A, Tran N Q, Ward J F and. Milligan J R (2004). Radioprotection of plasmid DNA by oligolysines. *Int. J. Radiat. Biol.* 80(9), 643–651.

5. Paul P, Unnikrishnan M K and Nagappa A N (2011). Phytochemicals as radioprotective agents-A review. *Indian J. Nat. Prod. Reso.* 2(2), 137-150.

6. Kumaran S P, Kutty B C, Chatterji A, Subrayan P P and Mishra K P (2007). Radioprotection against DNA damage by an extract of Indian green mussel, *Perna viridis* (L). *J. Environ. Pathol. Toxicol. Oncol.* 26(4), 263-72.

7. Zheng S, Newton GL, Gonick G, Fahey RC, Ward JF (1988). Radioprotection of DNA by thiols: Relationship between the net charge on a thiol and its ability to protect DNA. *Radiat Res.* 114,11–27.

8. Jamwal V S, Mishra S, Singh A, Kumar R (2014). Free radical scavenging and radioprotective activities of Hydroquinone *in vitro. J. Radioprot. Res.* 2(3),37-45.

Protocol No. 2.7: Determination of Radiation Induced Functional Impairment of Restriction Enzymes and its Protection by Radioprotective Compounds *In vitro* Assay

Assay Requirement

Restriction enzyme (ECOR1 or Bam H1 or Hind III), λ phage DNA, restriction buffer, agarose, ethidium bromide, incubator, UV- illuminator, gel documentation system.

2.7.1 Assay Procedure

2.7.1.1 To estimate the radiation induced functional impairment and its protection by radioprotective agent a modified assay of Daly *et al.* (2010) can be used.

2.7.1.2 Following sets of reaction mixture were formed:

   Set 1: 1.0 unit (5µl) of restriction enzyme (*i.e.* ECOR1 or Bam H1 or Hind III) was mixed with 5µl of test compound (stock solution 1.0 mg/ml) and incubated at 37°C for 30 minutes.

   Set 2: 1.0 unit (5µl) of either restriction enzyme (*i.e.* ECOR1 or Bam H1 or Hind III) without any test compound was incubated at 37°C for 30 minutes as control.

2.7.1.3 Followed by incubation, both reaction mixtures (set 1 and 2) were irradiated with different doses (800-1400Gy) of gamma radiation individually.

2.7.1.4 Unirradiated normal restriction endonuclease (*i.e.* ECOR1or Bam H1 or Hind III) was referred as positive control in the experiment.

2.7.1.5 Followed by irradiation, restriction digestion reaction was performed using both sets [irradiated in presence (Set 1) and absence (Set 2) of radioprotective compound] of endonucleases (*i.e.* ECOR1or Bam H1 or Hind III) and λ phage DNA(300ng) in 10x restriction buffer at 37°C for 60 minutes.

2.7.1.6 Upon incubation, entire reaction mixture was analyzed on 1 per cent agarose gel electrophoresis and stained with ethidium bromide.

2.7.1.7 To determine the protective efficacy of irradiated restriction endonucleases in presence and absence of test compound, restriction fragments of λ phage DNA appeared on agarose gel with both the sets were compared.

References

1. Mishra S, Gupta A K, Malhotra P, Singh P K, Pathak R, Gautam H K, Singh A, Kukreti S K, Javed S and Kumar R (2013). Protection against ionizing radiation induced oxidative damage to structural and functional proteins by semiquinone glucoside derivative isolated from radioresistant bacterium *Bacillus* sp. INM-1. *Current Bioechnol.* 3, 1-9.

2. Mishra S, Malhotra P, Gupta A K, Singh P K, Javed S and Kumar R (2013). A method for screening of radioprotiveection agents providing protection against gamma radiation induced functional impairment of DNA ligase. *Int J Radiat Biol.* DOI: 10.3109/09553002.2014.868613.

3. Saha A, Mandal P C and Bhattacharya S N (1995). Radiation-induced inactivation of enzymes. *Radiat Phys Chem.* 46, 123-45.

4. Daly M J, Gaidamakova E K, Matrosova V Y, *et al.* (2010). Small-molecule antioxidant proteome-shields in *Deinococcus radiodurans*. *PloS One.* 5, e12570.

5. Daly M J, Gaidamakova E K, Matrosova V Y, *et al.* (2007). Protein oxidation implicated as the primary determinant of bacterial radio resistance. *PLoS Biol.* 5, e92.

6. Roberts R J and Halford S E (1993). Type II restriction endonucleases. In: *Nucleases*, Linn SM, Lloyd RS, Roberts RJ, Eds. Cold Spring Harbour Laboratory Press, New York, pp. 35-88.

Protocol No. 2.8: Determination of Radioprotective Efficacy of Radioprotective Compound against Gamma Radiation Induced Functional Impairment of DNA Ligase

Assay Requirement

Genomic or plasmid DNA, restriction enzymes (any endonucleases), restriction buffer, T4 DNA ligase, ligation buffer, nuclease free water, agarose, TAE buffer (pH 8.2-8.4), UV illuminator, gamma irradiator and DNA gel electrophoresis unit with power supply.

2.8.1 Step1 Restriction Digestion of Genomic or Plasmid DNA

2.8.1.1 Genomic or plasmid DNA (200-300ng) was mixed with 1.0 unit of Hind III or any other restriction endonuclease in 10X restriction buffer and incubated for 1h at 37°C to complete the DNA digestion.

2.8.1.2 Digestion reaction was terminated by incubation of reaction mixture at 65°C for 10 minutes that deactivates the restriction enzymes.

2.8.1.3 Two parallel sets (set 1 and 2) of restriction digestion reaction were prepared separately.

2.8.2 Step 2 Inactivation of T4 DNA Ligase by Irradiation

2.8.2.1 T4 DNA ligase (1 unit equal to 200 weiss units) was taken in two individual sets, set A and set B.

2.8.2.2 Equal amount of test compound (natural or synthetic) under screening was added to set A, while set B (without test compound) was used as control.

2.8.2.3 Both the sets were incubated for 10-15 minutes at 25°C to allow compound to react with DNA ligase.

2.8.2.4 Both the sets were then irradiated with different doses of gamma radiation (>3000-7000Gy).

2.8.3 Step 3 Modified Ligation Reactions

2.8.3.1 One set of digested DNA (200-300ng) obtained from step A was mixed with DNA ligase irradiated with gamma radiation in absence of test compound (refer Set A, step 2).

2.8.3.2 While, second sets of digested DNA was mixed with T4 DNA ligase irradiated in presence of test compound (refer Set B, step 2). In brief, following sets of reaction mixture was formed:

Set A: Digested DNA (200-300ng) +1-5 unit of irradiated T4 DNA ligase + 4µl of 5X ligation buffer + nuclease free water (to make final volume up to 20µl).

Set B: Digested DNA (200-300ng in 2-3µl) + 1-5 unit of T4 DNA ligase + varied amount (1-5 mg) of natural/synthetic drug + 4µl of 5X ligation buffer + nuclease free water (to make the final volume to 20µl).

2.8.3.3 Both the sets (A and B) of reaction mixture incubated at 22°C for 20 minutes for ligation of restricted DNA.

2.8.3.4 Followed by ligation, reaction mixture was loaded on agarose (1 per cent) gel and electrophoresis performed at constant voltage of 45V-60V for 60-90 minutes in TAE buffer (pH 8.2-8.4) (1-6).

2.8.3.5 The ligation product of both the sets was analyzed under UV illuminator.

References

1. Mishra S, Gupta AK, Malhotra P, Singh PK, Pathak R, Gautam HK, Singh A, Kukreti SK, Javed S and Kumar R (2013). Protection against ionizing radiation induced oxidative damage to structural and functional proteins by semiquinoneglucoside derivative isolated from radioresistant bacterium *Bacillus* sp. INM-1. *Current Bioechnol.* 3, 1-9.

2. Mishra S, Malhotra P, Gupta A K, Singh P K, Javed S and Kumar R (2013). A method for screening of radioprotective action agents providing protection against gamma radiation induced functional impairment of DNA ligase. *Int J Radiat Biol.* DOI: 10.3109/09553002.2014.868613.

3. Saha A, Mandal P C and Bhattacharya S N (1995). Radiation-induced inactivation of enzymes. *Radiat Phys Chem.* 46, 123-45.

4. Daly M J, Gaidamakova E K, Matrosova V Y, *et al.* (2010). Small-molecule antioxidant proteome-shields in *Deinococcus radiodurans*. *PloS One.* 5, e12570.

5. Daly M J, Gaidamakova E K, Matrosova V Y, *et al.* (2007). Protein oxidation implicated as the primary determinant of bacterial radio resistance. *PLoS Biol.* 5, e92.

6. Roberts R J and Halford S E (1993). Type II restriction endonucleases. In: *Nucleases*, Linn SM, Lloyd RS, Roberts RJ, Eds. Cold Spring Harbour Laboratory Press, New York, pp. 35-88.

Section 3

Advance *In vitro* Methods of
Radioprotective Efficacy Evaluation

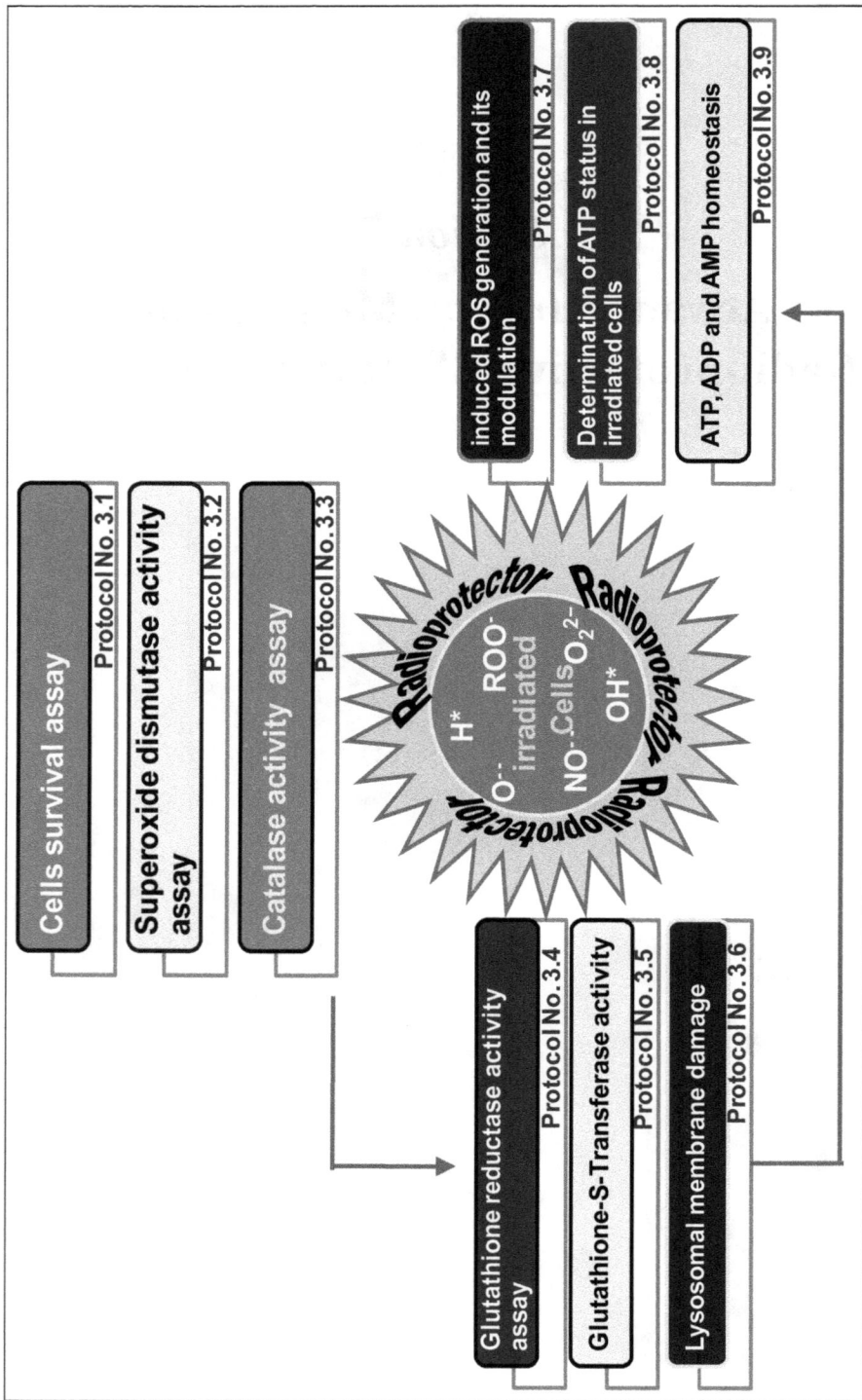

Figure 11: Schematic Summary of Advances *In vitro* Methods to Screen Radioprotective Drugs/Agents.

Protocol No 3.1: Evaluation of Survival Benefit to the Irradiated Cells Pretreated with Radioprotective Compound(s): Colony Forming Unite (CFU) Assay

Assay Requirement

Biosafety cabinet level-II, CO_2 incubator, plastic culture dishes, automated cell counter or heamocytometer, inverted microscope, MEM culture media, Fetal Bovine Serum (FBS), sterile micro-pipettes and tryphan blue.

Assay Procedure

3.1.1 Cell Culture

3.1.1.1 Cells from human or animals origin were cultured and maintained in CO_2 incubator as mono-layer culture in minimal essential medium (MEM) supplemented with 10 per cent FBS, 100 units/ml of penicillin and 100µg/ml of streptomycin.

3.1.1.2 The pH of the culture medium was fixed at pH 7.4. Cells were incubated in 25 cm² culture flasks at 37 °C in a humidified (95 per cent) environment with 5 per cent CO_2 concentration.

3.1.1.3 Cells were sub-cultured twice a week. All experiments were performed when cells reached about 70 per cent confluency.

3.1.2 Cells Harvesting and Counting

3.1.2.1 Cultured cells were harvested using 0.25 per cent Trypsin in Hank's balanced salt solution (HBSS).

3.1.2.2 In brief, culture medium (MEM) was decanted from the culture dishes or flasks.

3.1.2.3 1.0 ml of chilled trypsin (0.25 per cent) was poured into the culture dishes/flask (T25) and incubated for 30 seconds to 1 minute at 37 °C temperature.

3.1.2.4 Followed by incubation, excess trypsin solution was decanted.

3.1.2.5 The culture dishes/flasks were incubated at 37°C temperature again till rolling off of the cells started.

3.1.2.6 De-attached cells were harvested in a sterile 2.0 ml microtube.

3.1.2.7 Cells were counted with the help of cytometer or automatic cell counter.

3.1.2.8 Proper dilutions of cell suspension was made by adding complete MEM medium to the cell suspension and fixed number of cells was seeded in separate culture dishes or flasks.

3.1.3 Colony Forming Unit Assay

3.1.3.1 Harvested cells were pipetted gently several times to make a single cell suspension and counted with the help of cell counter.

3.1.3.2 Viability of cells was tested by trypan blue dye (0.1 per cent) exclusion test.

3.1.3.3 Nearly 300 cells were seeded in 12 culture dishes (60 mm diameter) and incubated for 12 h at 37 °C with 95 per cent humidity and 5 per cent CO_2 concentration.

3.1.3.4 The culture dishes were grouped into following four groups in triplicate

    i. Control (dishes n =3)

    ii. Irradiated cells (dishes n =3)

    iii. Cells treated with radioprotective compound (dishes n =3)

    iv. Irradiated cells pretreated with radioprotective compound (dishes n = 3)

3.1.3.5 Attached cells were subjected to respective treatments according to the groups mentioned above.

3.1.3.6 Followed by different treatments, culture medium from all the dishes was replaced with the fresh complete culture medium.

3.1.3.7 The cells were further incubated at 37°C with 5 per cent CO_2 in a 95 per cent humidly at least two weeks.

3.1.3.8 To avoid the nutritional deficiency, culture medium of the cells was changed after every 48 h.

3.1.3.9 Colonies appeared nearly after two weeks of incubation was counted and cell survival fractions calculated as per cent survival in all the treated groups.

3.1.3.10 The per cent survival of irradiated cells was compared with the survival (per cent) of irradiated cells pretreated with radioprotective compound (1-5).

References

1.   http://www.rndsystems.com/literature-CFC.aspx.

2.   Pochampally R (2008). Colony forming unit assays for MSCs. *Methods Mol Biol.* 449,83 91. doi: 10.1007/978-1-60327-169-1-6.

3.   Avraham H K, Lee BC and Avraham S (2005). Colony-forming unit assay: Methods and implications. *Encyclopedic Reference of Immunotoxicology*, pp. 144-147.

4. Penfornis P and Pochampally R (2011). Isolation and expansion of mesenchymal stem cells/multipotential stromal cells from human bone marrow. *Methods Mol Biol*. 698,11-21. doi: 10.1007/978-1-60761-999-4-2.

5. Alexey B (2013). Standardization of colony-forming unit assay for hematopoietic products potency. http://stemcellassays.com/2013/03/cfu-standardization.

Protocol No 3.2: Superoxide Dismutase Activity Evaluation in Irradiated Cells and its Modulations by Radioprotective Compounds Pretreatment using *In vitro* Cell Culture Method

Assay Requirement

Biosafety cabinet level-II, CO_2 incubator, plastic culture dishes, automated cell counter or heamocytometer, inverted microscope, centrifuge, MEM culture media, Fetal Bovine Serum (FBS), sterile micro-pipettes and tryphan blue, PBS buffer (pH 7.4), NP-40, protease inhibitor, sucrose buffer, high salt buffer, bradford reagent, acrylamide, Bis-acrylamide, Tris-HCl, glycerol, α-mercaptoethanol, bromophenol blue, Tris- EDTA buffer, Tris-glycine, Nitrobluetetazolium (NBT), Riboflavin, TEMED and densitometry software.

Assay Procedure

3.2.1 Cell Culture and Experimental Groups Formation

3.2.1.1 Equal number (5.0 millions) of cells was grown *in vitro* culture in culture flask at 37°C temperature with 95 per cent humidity and 5 per cent CO_2 concentration for 24-48h (depending on the doubling time) in the following experimental groups:

    i. Untreated Control cells

    ii. Irradiated cells (cells exposed with desired dose of gamma radiation; (1-4Gy)

    iii. Cells treated with radioprotective compound

    iv. Irradiated cells pretreated with radioprotective compound.

3.2.1.1 Individual group of cells was subjected to respective drug and gamma radiation treatment and incubated in CO_2(5 per cent) incubator at 37°C with 95 per cent humidity.

3.2.1.2 Cells were trypsinized and harvested as described in previous protocol (No.3.1.2)

3.2.1.3 Harvested cells were centrifuged at 2000xg for 5 minutes and cell pellet dissolved in 500 µl of ice cold sterile PBS buffer (pH 7.4).

3.2.2 Extraction of Cytoplasmic and Nuclear Protein from Cultured Cells

3.2.2.1 Cells suspension was centrifuged at 2000xg for 5 minutes and supernatant discarded.

3.2.2.1 Cells pellet was dissolved in 1ml of PBS (pH 7.4) and again centrifuged at 2000xg for 5 minutes.

3.2.2.2 Cells were re-suspended in 200µl of sucrose buffer with NP-40 and protease inhibitor by gently pipetting and incubated on the ice for 5 minutes.

3.2.2.3 Nuclei of the cells were pelleted out by centrifugation at 15000xg for 15 minutes at 4°C and transfer the supernatant into new sterile tubes.

3.2.2.4 The clear supernatant was stored at 4 °C for analysis of cytoplasmic fractions.

3.2.2.5 Cells pellet was resuspended in sucrose buffer without NP-40 and centrifuged at 15000xg for 5 minutes. Supernatant was discarded.

3.2.2.6 The pellet was resuspended in 200µl low salt buffer and then 200µl high salt buffers for 5 minutes.

3.2.2.7 Cells mixture was incubated on cold rotator platform for 20 minutes and centrifuged at 15000xg for 10 minutes at 4°C.

3.2.2.8 The clear supernatant obtained was stored at 4°C for analysis of nuclear fraction.

3.2.3 Estimation of Protein Content in Cytoplasmic and Nuclear Fraction

3.2.3.1 Total soluble protein contents in the different cytoplasmic and nuclear fractions were estimated by Bradford (1976) method.

3.2.3.2 Briefly, 10µl of the sample was mixed with 90µl of double distilled water.

3.2.3.3 After thorough mixing, Bradford reagent (1.0ml) added and mixture was incubated for 10 minutes at room temperature.

3.2.3.4 Absorbance was recorded at room temperature at 595nm using microplate spectrophotometer.

3.2.3.5 The amount of protein present in particular fraction was quantified using BSA standard prepared similarly.

3.2.3.6 Calculated amount (10µg) of protein samples was used for SOD expression analysis using native poly-acrylamide gel electrophoresis.

3.2.4 Native PAGE Analysis

3.2.4.1 10 per cent polyacrylamide (without SDS) resolving and 4 per cent stacking gels of 0.75 mm thickness was prepared.

3.2.4.2 Protein (equal to 10µg of protein in each sample) extracted from the all treatment groups of the cells was mixed with the sample buffer [(0.0625 M Tris-HCl, pH 6.8; 5 per cent (v/v) glycerol; 2 per cent (v/v) α-mercaptoethanol; 0.01 per cent (w/v) bromophenol blue] without heating or boiling.

3.2.4.3 A pre-electrophoretic run for half an hour was performed using Tris-EDTA (pH 7.2) running buffer to remove oxidants from the gel which may interfere with enzyme activity.

3.2.4.4 Followed by pre-electrophoretic run, protein samples (preferably 8-10µg) was loaded on the gel. Electrophoresis was carried out using Tris-glycine running buffer and a constant voltage (stacking at 60V, resolving at 100V).

3.2.4.5 Followed by electrophoresis, gel was placed in the dark with staining solution-1(1.23mM Nitro blue tetazolium) for 25 minutes.

3.2.4.6 After that, gel was stained with staining solution-2 (0.028 mm Riboflavin and 28mM TEMED). Gels were then allowed to exposed with direct light to activate photoreaction.

3.2.4.7 Transparent bands against the dark background were appeared on the gels (1-6).

3.2.4.8 Quantification of the bands was carried out using densitometric analysis software.

References

1. Odén P C, Karlsson G and Einarsson R (1992). Demonstration of superoxide dismutase enzymes in extracts of pollen and anther of *Zea mays* and in two related products, Baxtin® and Polbax®. Grana. 31, 76-80.

2. Chen C N and Pan S M (1996). Assay of superoxide dismutase activity by combining electrophoresis and densitometry. *Bot. Bull. Acad. Sin.* 37, 107-111.

3. Beauchamp C and Fridovich I (1971). Superoxide dismutase: Improved assays and an assay applicable to acrylamide gels. *Anal. Biochem.* 44, 276-287.

4. Janknegt P J, Rijstenbil J W, Willem H. Poll V, Gechev T S and Buma A G J (2007). A comparison of quantitative and qualitative superoxide dismutase assays for application to low temperature microalgae. *J. Photochem. Photobiol. B: Biology.* 87, 218–226.

5. Kuo WY, Huang C H and Jinn T L (2013).Cellular extract preparation for superoxide dismutase (SOD) activity assay. *Bio Protocol.* 7-5. http://www.bio-protocol.org/e811.

6. Kuo W Y, Huang C H, Liu A C, Cheng C P, Li S H, Chang W C, Weiss C, Azem A and Jinn T L (2013). CHAPERONIN 20 mediates iron superoxide dismutase (FeSOD) activity independent of its co-chaperonin role in *Arabidopsis* chloroplasts. *New Phytol.* 197(1), 99-110.

Protocol No. 3.3: Catalase Activity Evaluation in Irradiated Cells and its Modulations by Radioprotective Compounds Pretreatment using *In vitro* Cell Culture Method

Assay Requirement

Biosafety cabinet level-II, CO_2 incubator, plastic culture dishes, automated cell counter or heamocytometer, inverted microscope, centrifuge, MEM culture media, Fetal Bovine Serum (FBS), sterile micro-pipettes and tryphan blue, PBS buffer (pH 7.4), NP-40, protease inhibitor, sucrose buffer, high salt buffer, Bradford reagent, acrylamide, Bis-acrylamide, Tris-HCl, glycerol,α-mercaptoethanol, bromophenol blue, Tris- EDTA buffer, Tris-glycine, H_2O_2 ferric chloride, potassium ferricyanide and densitometry software.

Assay Procedure

3.3.1 Cell Culture and Experimental Groups Formation

3.3.1.1 Equal number (5.0 millions) of cells were grown *in vitro* culture in culture flask at 37°C temperature with 95 per cent humidity and 5 per cent CO_2 concentration for 24-48h (depending on the doubling time) in the following experimental groups:

  i. Untreated control cells

  ii. Irradiated cells (cells exposed with desired dose of gamma radiation; 2-4 Gy)

  iii. Cells treated with radioprotective compounds

  iv. Cells treated with radioprotective compounds before gamma irradiation.

3.2.1.2 Individual group of cells was subjected to respective drug and gamma radiation treatments. Now cells were trypsinized and harvested as described in the protocol No.3.1.2).

3.3.1.3 Harvested cells were centrifuged at 2000xg for 5 minutes and cell pellet dissolved in 500µl of ice cold sterile PBS buffer (pH 7.4).

3.3.2 Extraction of Cytoplasmic and Nuclear Protein from Cultured Cells

Please Refer Protocol No. 3.2.2

3.3.2.1 In brief, cells pellet was dissolved in 1.0 ml of PBS (pH 7.4) and centrifuged at 2000xg for 5 minutes.

3.3.2.2 Cells were re-suspended in 200µl of sucrose buffer with NP-40 and protease inhibitor by gentle pipetting and cells suspension incubated on ice for 5 minutes.

3.3.2.3 Nuclei of the cells were pelleted out by centrifugation at 15000 xg for 15 minutes at 4°C and transfer the supernatant into new tubes.

3.3.2.4 The clear supernatant was stored at 4 °C for analysis of cytoplasmic fractions.

3.3.2.5 Cell pellet was re-suspended in sucrose buffer without NP-40 and centrifuged at 15000xg for 5 minutes and supernatant discarded.

3.3.2.6 The pellet was re-suspended in 200 µl low salt buffer and then 200 µl high salt buffers for 5 minutes.

3.3.2.7 Cells suspension was incubated on cold rotator platform for 20 minutes and centrifuged at 15000xg for 10 minutes at 4°C.

3.3.2.8 The clear supernatant obtained was stored at 4°C for analysis of nuclear fraction.

3.3.3 Estimation of Protein Content in Cytoplasmic and Nuclear Fractions

3.3.3.1 Total soluble protein contents in the different cytoplasmic and nuclear fractions were estimated by Bradford (1976) method.

3.3.3.2 Briefly, 10 µl of the sample was mixed with 90 µl of double distilled water.

3.3.3.3 After thorough mixing, 1.0 ml of Bradford reagent was added and the absorbance was recorded at room temperature at 595nm using microplate spectrophotometer.

3.3.3.4 The amount of protein present in particular fraction was quantified using BSA standard prepared simultaneously. Calculated amount (10µg) of protein samples was used for catalase expression analysis using native PAGE.

3.3.4 Native Gel Analysis

3.3.4.1 A non-denaturing polyacrylamide gel (8 per cent) was prepared as described in previous protocols.

3.3.4.2 A pre-electrophoretic run of 30 minutes was performed using Tris- EDTA as running buffer to remove oxidants which might alter enzyme activity.

3.3.4.3 After pre-electrophoretic run, protein samples were loaded and separated by electrophoresis in the presence of Tris-glycine buffer.

3.3.4.4 The gel was stained with 0.4 per cent H_2O_2 for 5 min.

3.3.4.5 The gel was soaked in 1 per cent ferric chloride and 1 per cent potassium ferricyanide for 1 minutesand then the stain removed (1-5).

3.3.4.6 White coloured bands appeared on the dark background were analyzed using quantifiable densitometric software.

References

1. Sahu S, Das P, Ray M, Sabat S C (2010). Osmolyte modulated enhanced rice leaf catalase activity under salt-stress. *Adv. Biosci. Biotechnol.* 1, 39-46.doi:10.4236/abb.2010.11006.

2. Weydert C J and Cullen J J (2010). Measurement of superoxide dismutase, catalase and glutathione peroxidase in cultured cells and tissue. Nat. Protocol. 5(1), 57-66. Doi: 10.1038/nprot.2009-197.

3. Petrova V Y, Rasheva T V and Kujumdzieva A V (2002). Catalase enzyme in mitochondria of *Saccharomyces cerevisiae*. *EJB Electronic J. Biotechnol.* DOI: 10.2225/vol5-issue1-fulltext-6.

4. Scherer M, Wei H, Liese R and Fischer R (2002). *Aspergillus nidulans* catalase-peroxidase gene (*cpeA*) is transcriptionally induced during sexual development through the transcription factor StuA. *Eukaryotic Cell.* 1(5),725-735.

5. Garcia M X U, Foote C, Es S V and Devreotes P N (2000). Differential developmental expression and cell type specificity of *Dictyostelium* catalases and their response to oxidative stress and UV-light. *Biochimica. Biophysica. Acta.* 1492, 295-310.

Protocol No. 3.4: Glutathione Reductase Activity Evaluation in Irradiated Cells and its Modulations by Radioprotective Compound Pretreatment Using *In vitro* Cell Culture Method

Assay Requirement

Biosafety cabinet level-II, CO_2 incubator, plastic culture dishes, automated cell counter or heamocytometer, inverted microscope, centrifuge, MEM culture media, Fetal Bovine Serum (FBS), sterile micro-pipettes and tryphan blue, PBS buffer (pH 7.4), NP-40, protease inhibitor, sucrose buffer, high salt buffer, Bradford reagent, acrylamide, Bis-acrylamide, Tris-HCl, glycerol, α-mercaptoethanol, bromophenol blue, Tris-EDTA buffer, Tris-glycine, DCIP, GSSG, NADPH, MTT and densitometry software.

3.4.1 Assay Procedure

3.4.1.1 Equal number (5.0 millions) of cells were grown *in vitro* culture in culture flask at 37°C temperature with 95 per cent humidity and 5 per cent CO_2 concentration for 24-48h (depending on the doubling time) in the following experimental groups:

    i. Untreated control

    ii. Irradiated cells [cells exposed with desired dose (2-4 Gy) of gamma radiation]

    iii. Cells treated with desired concentration of radioprotective compounds

    iv. Cells treated with desired concentrations of radioprotective compounds before gamma irradiation.

3.4.1.2 After different treatments, cells were allowed to incubate for desired time period (6-48h).

3.4.1.3 Followed by incubation, cells were harvested by trypsinization as described in the previous protocol No.3.1.2).

3.4.1.4 Harvested cells were centrifuged at 2000xg for 5 minutes and cell pellet dissolved in 500 µl of ice cold sterile PBS buffer (pH 7.4).

3.4.2 Extraction of Cytoplasmic and Buclear Protein from Cultured Cells

Please refer protocol No.3.2.2

3.4.2.1 Cells grown in different treatments groups were harvested at different time points and centrifuged at 2000xg for 5 minutes. Supernatant obtained was discarded.

3.4.2.2 Cells pellet was dissolved in 1.0 ml of PBS (pH 7.4) and again centrifuged at 2000xg for 5 minutes.

3.4.2.3 Cells were re-suspended in ice cold 200µl of sucrose buffer (containing NP-40 and protease inhibitor) for 5 minutes.

3.4.2.4 Nuclei of the cells were pelleted out by centrifugation at 15000xg for 15 minutes at 4°C and transfer the supernatant into new tubes.

3.4.2.5 The clear supernatant was stored at 4 °C for analysis of cytoplasmic fractions.

3.4.2.6 Cells pellet was re-suspended in sucrose buffer without NP-40 and centrifuge at 15000xg for 5 minutes and supernatant discarded.

3.4.2.7 The pellet was re-suspended in 200 µl low salt buffer and then 200µl high salt buffers for 5 minutes.

3.4.2.8 Cells suspension was incubated on cold rotator platform for 20 minutes and centrifuged at 15000xg for 10 minutes at 4°C.

3.4.2.9 The clear supernatant obtained was stored at 4°C for analysis of nuclear fraction.

3.4.3 Estimation of Protein Content in Cytoplasmic and Nuclear Fractions

3.4.3.1 Total soluble protein contents in the different cytoplasmic and nuclear fractions were estimated by Bradford (1976) method.

3.4.3.2 Briefly, 10 µl of the sample was mixed with 90 µl of water.

3.4.3.3 After thorough mixing, Bradford reagent (1.0 ml) was added and reaction mixture incubated for 10 minutes at room temperature.

3.4.3.4 Absorbance was recorded at room temperature at 595nm using microplate spectrophotometer.

3.4.3.5 The amount of the protein present in the particular fraction was quantified using BSA standard.

3.4.3.6 Calculated amount (10µg) of protein samples was used for glutathione reductase expression analysis on native Polyacrylamide gel electrophoresis.

3.4.4 Native Polyacrylamide Gel Electrophoresis

3.4.4.1 A 10 per cent non-denaturing polyacrylamide gel was prepared.

3.4.4.2 A pre-electrophoretic run of 30 minutes was carried out using Tris- EDTA running buffer to remove oxidants which might alter enzyme activity.

3.4.4.3 Followed by pre-electrophoretic run, equal amount of protein (8-10 µg) was loaded on polyacrylamide gel and electrophoresis (as described in previous section) performed.

3.4.4.4 The gel was separated out from the electrophoresis module and kept in the dark with the staining solution (4.24mg DCIP, 52.1mg GSSG, 8.0mg NADPH and11.4mg MTT) prepared in 25ml of 250 mMTris (pH 8.0).

3.4.4.5 The gel was observed for appearance of purple bands against transparent background (1-9).

3.4.4.6 Quantitative analysis of the purple colored bands was carried out using densitometric software.

References

1.   Hou W C, Liang H J, Wang C C, Liu D Z (2004). Detection of glutathione reductase after electrophoresis on native or sodium dodecyl sulfate polyacrylamide gels. *Electrophoresis*. 25(17), 2926-2931.

2.   Yannarelli G G, Ferna´ndez-Alvarez AJ, Santa-Cruz D M and Tomaro M L (2007).Glutathione reductase activity and isoforms in leaves and roots of wheat plants subjected to cadmium stress. *Phytochemistry*. 68, 505–512.

3.   Dewir Y H, Chakrabarty D, Ali MB, Hahn EJ and Paek KY (2005). Lipid peroxidation and antioxidant enzyme activities of *Euphorbia millii* hyperhydric shoots. *Environ. Experi. Botany*. Doi:10.1016/j.envexpbot.2005.06.019.

4.   Sumugat M R S (2004). Glutathione dynamics in *arabidopsis* seed development and germination. *Thesis*. 1-98. http://scholar.lib.vt.edu/theses/available/etd-12232004-103334/unrestricted/THESIS.pdf.

5.   Calbert I and Mannervik B (1985). Glutathione reductase. *Meth. Enzymol*. 113, 484–490.

6.   Lascano HR, Go´mez LD, Casano LM and Trippi VS (1998). Changes in glutathione reductase activity and protein content in wheat leaves and chloroplasts exposed to photooxidative stress. *Plant Physiol. Biochem*. 36, 321–329.

7.   Madamanchi NR, Anderson JV, Alscher RG, Cramer CL and Hess JL (1992). Purification of multiple forms of glutathione reductase from pea (*Pisum sativum* L.) seedlings and enzyme levels in ozone-fumigated pea leaves. *Plant Physiol*. 100, 138–145.

8.   Edwards EA, Enard C, Creissen GP and Mullineaaux PM (1994). Synthesis and properties of glutathione reductase in stressed peas. *Planta*. 192, 137–143.

9.   Anderson JV and Davis DG (2004). Abiotic stress alters transcript profiles and activity of glutathione S-transferase, glutathione peroxidase and glutathione reductase in *Euphorbia esula*. *Physiol. Plant*. 120, 421–433.

Protocol No. 3.5: Glutathione-S-Transferase Activity Evaluation in Irradiated Cells and its Modulations by Radioprotective Compound Pretreatment using *In vitro* Cell Culture Method

Assay Requirement

Biosafety cabinet level-II, CO_2 incubator, plastic culture dishes, automated cell counter or heamocytometer, inverted microscope, centrifuge, MEM culture media, Fetal Bovine Serum (FBS), sterile micro-pipettes and tryphan blue, PBS buffer(pH7.4), NP-40, protease inhibitor, sucrose buffer, high salt buffer, Bradford reagent, acrylamide, Bis-acrylamide, Tris-HCl, glycerol, a-mercaptoethanol, bromophenol blue, Tris-EDTA buffer, Tris-glycine, Trypsin, DCIP, GSSG, NADPH, MTT and densitometry software.

3.5.1 Assay Procedure

3.5.1.1 Equal number (5.0 millions) of cells were grown *in vitro* culture in culture flask at 37°C temperature with 95 per cent humidity for 24-48 h (depending on the doubling time) in the following experimental groups:

   i. Untreated control

   ii. Irradiated cells (cells exposed with desired dose (2-4 Gy) of gamma radiation)

   iii. Cells treated with desired concentration of radioprotective compounds

   iv. Cells treated with desired concentrations of radioprotective compounds before gamma irradiation.

3.5.1.2 After different treatments, cells were allowed to incubate for desired time period (6-48h).

3.5.1.3 Followed by incubation cells were harvested by trypsinization as described in the previous protocol No.3.1.2)

3.5.1.4 Harvested cells were centrifuged at 2000 x g for 5 minutes and cell pellet dissolved in 500 µl of ice cold sterile PBS buffer (pH 7.4).

3.5.2 Extraction of Cytoplasmic and Nuclear Protein from Cultured Cells

3.5.2.1 Cells grown in different treatments groups were harvested at different time points and centrifuged at 2000xg for 5 minutes. Supernatant obtained was discarded.

3.5.2.2 Cells pellet was dissolved in 1.0 ml of PBS (pH 7.4) and again centrifuged at 2000xg for 5 minutes.

3.5.2.3 Cells were re-suspended in ice cold 200μl of sucrose buffer (containing NP-40 and protease inhibitor) for 5 minutes.

3.5.2.4 Nuclei of the cells were pelleted out by centrifugation at 15000 xg for 15 minutes at 4°C and transfer the supernatant into new tubes.

3.5.2.5 The clear supernatant was stored at 4 °C for analysis of cytoplasmic fractions.

3.5.2.6 Cells pellet was re-suspended in sucrose buffer without NP-40 and centrifuge at 15000 xg for 5 minutes and supernatant discarded.

3.5.2.7 The pellet was re-suspended in 200 μl low salt buffer and then 200 μl high salt buffers for 5 minutes.

3.5.2.8 Cells suspension was incubated on cold rotator platform for 20 minutes and centrifuged at 15000xg for 10 minutes at 4°C.

3.5.2.9 The clear supernatant obtained was stored at 4°C for analysis of nuclear fraction.

3.5.3 Estimation of Protein Content in Cytoplasmic and Nuclear Fractions

3.5.3.1 Total soluble protein contents in the different cytoplasmic and nuclear fractions were estimated by Bradford (1976) method.

3.5.3.2 Briefly, 10μl of the sample was mixed with 90 μl of water.

3.5.3.3 After thorough mixing, Bradford reagent (1.0 ml) was added and reaction mixture incubated for 10 minutes at room temperature.

3.5.3.4 Absorbance was recorded at room temperature at 595nm using microplate spectrophotometer.

3.5.3.5 The amount of the protein present in the particular fraction was quantified using BSA standard.

3.5.3.6 Calculated amount (10μg) of protein samples was used for glutathione-S-Transferase expression analysis on native polyacrylamide gel electrophoresis.

3.5.4 Native PAGE Analysis

3.5.4.1 A non-denaturing polyacrylamide gel (12 per cent) was prepared.

3.5.4.2 A pre-electrophoretic 30 minutes run was performed using Tris-EDTA running buffer to remove oxidants which might alter enzyme activity.

3.5.4.3 Equal amount of protein was loaded in each well and electrophoresis performed as described in previous protocol.

3.5.4.4 The gel was kept in the dark and stained with the solution containing 4.24mg DCIP, 52.1mg GSSG, 8.0mg NADPH and11.4mg MTT in 25ml of 250 mMTris (pH 8.0).

3.5.4.4 The gel was observed for purple bands appeared against transparent background (1-7).

References

1. Anderson J V and Davis D G (2004). Abiotic stress alters transcript profiles and activity of glutathione S-transferase, glutathione peroxidase, and glutathione reductase in *Euphorbia esula. Physiol. Plant.* 120, 421–433.

2. Prabhu K S, Reddy P V, GumprichtE, Hildenbrandt G R, Scholz RW, Sordillo L M and Reddy C C (2001). Microsomal glutathione S-transferase A1-1 with glutathione peroxidase activity from sheep liver: molecular cloning, expression and characterization. *Biochem. J.* 360, 345-354.

3. Hayes J D and Pulford D J (1995). The glutathione S-transferase super gene family : regulation of GST and the contribution of the isozymes to cancer chemoprotection and drug resistance. *Crit. Rev. Biochem. Mol. Biol.* 30, 445-600.

4. Mosialou E, Piemonte F, Andersson C, Vos-Ria M E, Van-Bladeren P J and Morgenstern R (1995). Microsomal glutathione transferase: lipid-derived substrates and lipid dependence. *Arch. Biochem. Biophys.* 320, 210-216.

5. Mannervik B, Berhane K, Bjo$rnestedt R, Board P G, Jones T A, Kolm R H, Olin B, Sinning I, Sroga G E, Stenberg G *et al.* (1993). Structural and functional characterization of the binding sites for glutathione (G-site) and the hydrophobic electrophilic substrate (H-site) in glutathione transferases. in *Structure and Function of Glutathione Transferases* (Tew K D, Pickett C B, Mantle T J, Mannervik Band Hayes J D, eds.), pp. 29-38, CRC Press, Boca Raton.

6. Morgenstern R, Guthenberg C and DePierre J W (1982). Microsomal glutathione S-transferase. Purification, initial characterization and demonstration that it is not identical to the cytosolic glutathione S-transferases A, B and C. *Eur. J. Biochem.* 128L 243-248.

7. Zibaee A, Bandani R, Haghani S and Zibaee A (2009). Partial characterization of glutathione-S-transferase in two populations of the sunn pest, *Eurygaster integriceps* puton(heteroptera: scutellaridae). *Mun. Ent. Zool.* 4(2), 564-571.

Protocol No. 3.6: Evaluation of Radiation Induced Lysosomal Membrane Damage and its Prevention by Radioprotective Compound: Lysosomal Membrane Stability Assay (LMSA)

Assay Requirement

Acrydine orange, DMEM culture media, 60mm culture discs, CO_2 incubator, gamma irradiator and flow cytometer equipped with green and red laser or fluorescence spectrometer.

Assay Procedure

3.6.1 Acrydine Orange (AO) Relocation Assay

3.6.1.1 Cells were grown in 60mm culture discs using standard cell culture procedures and divided into following four groups of experimental sets and incubated for 24h at 37 °C, 5 per cent CO_2 and 95 per cent humidity:

Set A: Control cells (non-irradiated and non-drug treated) stained with Acrydine orange (5mg/ml) for 30 minutes.

Set B: Cells treated with different concentrations of compound under evaluation (Radioprotector) in fresh medium.

Set C: Cell irradiated with gamma radiation (2-4 Gy) in absence of radioprotective compound under evaluation in fresh medium.

Set D: Cells irradiated with gamma radiation (2-4 Gy) in presence of radioprotective compound under evaluation in fresh medium.

3.6.1.2 Followed by different treatments the cells were incubated for 2-6 h and then stained with to 5 µg/ml Acrydine orange for 30 minutes at room temperature.

3.6.1.3 Changes in green fluorescence were determined by flow cytometry/ fluorescence spectrometer using excitation 502nm and emission 525 nm wavelengths (1-5).

References

1. Denamur S, Tyteca D, Marchand-Brynaert J, Van-Bambeke F, Tulkens P M, Courtoy P J, Leclercq M P M (2011). Role of oxidative stress in lysosomal membrane permeabilization and apoptosis induced by gentamicin, an aminoglycoside antibiotic. *Free Rad. Biol. Med.* 30(2011) xxx–xxx.

2. Zhao M, Eaton J W, Brunk UT (2000). Protection against oxidant-mediated lysosomal rupture: a new anti-apoptotic activity of Bcl-2? *FEBS Letter* 485, 104-108.

3. Yin L, Stearns R and Gonza'lez-Flecha B (2005). Lysosomal and mitochondrial pathways in H_2O_2-induced apoptosis of alveolar type II cells. *J. Cellu. Biochem.* 94: 433-445.

4. Zhao M, Antunes F, Eaton J W and Brunk UT(2003). Lysosomal enzymes promote mitochondrial oxidant production, cytochrome C release, and apoptosis. *Eur. J. Biochem.* 270: 3778–3786.

5. Chen YR, Wang X, Templeton D and Davis RJ, Th T (1996). The role of c-Jun N-terminal kinase (JNK) in apoptosis induced by ultraviolet C and gamma radiation. Duration of JNK activation may determine cell death and proliferation. *J. Biol. Chem.* 271: 31929–31936.

Protocol No. 3.7: Determination of Radiation Induced ROS Generation and its Modulation by Flow Cytometry or Time-Lapse Fluorescence Microscopy or Spectrometry

Assay Principle

The generation of reactive oxygen species (ROS) by gamma irradiation is typically observed in several cells and organs. The effect of ROS generation depends on its subcellular localization, concentration and their types. Dihydrorhodamine 123 (MW 346.4 Da) and CM-H$_2$DCFDA (MW 577.8 Da) are commonly used ROS indicators. Rhodamine123 is an uncharged and non-fluorescent ROS indicator that can passively diffuse across membranes, where it oxidized to cationic rhodamine 123, which localizes in the mitochondria and exhibits green fluorescence. CM-H$_2$DCFDA is a chloromethyl derivative of H$_2$DCFDA, but it is much better retained in live cells than H$_2$DCFDA. CM-H$_2$DCFDA passively diffuses into cells, where its acetate groups are cleaved by intracellular esterases and its thiol-reactive chloromethyl group reacts with intracellular glutathione and other thiols. Subsequent oxidation yields a fluorescent adduct that is trapped inside the cell, which facilitates long-term time lapse studies.

Assay Requirement

Rhodamine, Dihydrorhodamine 123, Propidium Iodide, polypropylene round bottom tube, flow cytometer/time-lapse fluorescence spectrophotometer or fluorescence microscope.

3.7.1 Assay Procedure

3.7.1.1 After seeding the cells and inducing cell death using gamma radiation in presence and absence of radioprotective drug add dihydrorhodamine 123 to the cells (1μM final).

3.7.1.2 Cell samples from suspension cultures were analyzed at regular intervals on the flow cytometer. For that a volume of 270 μl of cells suspension from the 24-well plate was is transferred to a 5-ml polypropylene round bottom tube, and 30 μl of cell impermeant probe (1-5), *e.g.* PI was added from a 10x solution (30 μM).

3.7.1.3 A synchronic cellular response in ROS generation results in an overall increase of the mean ROS detected, whereas an asynchronic response is detected by an increase in the amount of cells showing increased ROS levels (*e.g.* >200 per cent increase).

References

1. Berghe T V, Grootjans S, Goossens V, Dondelinger Y, Krysko D V, Takahashi N, Vandenabeele P (2013). Determination of apoptotic and necrotic cell death *in vitro* and *in vivo*. *Methods* 61, 117–129.

2. Ameziane-El-Hassani R and Dupuy C (2013). Detection of intracellular reactive oxygen species (CM-H2DCFDA). *Bio Protocol*, 3(1), 1-5. Doi: 10.3791/3357.

3. Yazdani M, Paulsen R E, Gjøen T, Hylland K (2014). Reactive oxygen species and cytotoxicity in rainbow trout hepatocytes: Effects of medium and incubation time. *Bull Environ. Contam. Toxicol.* DOI: 10.1007/s00128-014-1433-0.

4. Yang Y, Yang D, Yang D, Jia R, Ding G (2014). Role of reactive oxygen species-mediated endoplasmic reticulum stress in contrast-induced renal tubular cell apoptosis. *Nephron. Exp. Nephrol.* 128, 30-36.

5. Hsieh CY, Chen C L, Yang K C, TMa C, Choi P C and Lin C F (2015). Detection of reactive oxygen species during the cell cycle under normal culture conditions using a modified fixed-sample staining method. *J Immunoassay Immunochem.* 36(2), 149-6.

Protocol No. 3.8: Determination of ATP Status in Irradiated Cells and its Modulation by Radioprotective Compound Pretreatment Using Luciferin Luciferase System

Principle of the Assay

The reactions catalyzed by luciferase are:

i. Luciferase+ luciferin + ATP=luciferase–luciferyl–AMP + PPi

ii. Luciferase–luciferyl–AMP + O_2 = Luciferase +oxyluciferin + AMP + CO_2 + energy (hv)

The reaction produces a flash of yellow-green light, with a peak emission at 560nm, the intensity of which is proportional to the amount of substrates in the reaction mixture. To obtain a kinetic measurement of ATP synthesis, replicate tests could be performed on replicate samples incubated with appropriate substrates and arrested by lysis, generally in perchloric acid, at different time points.

Assay Requirement

Culture dishes, Dulbecco's modified Eagle's medium (DMEM), glucose, glutamine, sodium pyruvate, fetal bovine serum, trypsin, phosphate-buffered saline, digitonin, P^5-di(adenosine) pentaphosphate, malate, pyruvate, succinate, rotenone, luciferin, luciferase, Tris–acetate pH 7.75, ADP, oligomycin, centrifuge, luminometer, hemocytometer, gamma irradiator.

3.8.1 Assay Procedure

3.8.1.1 Cell culture

3.8.1.1.1 Cells were grown in 150 mm culture dishes in Dulbecco's modified Eagle's medium (DMEM) containing high levels of glucose (4.5 mg/ml), 2 mM L-glutamine, 110 mg/liter sodium pyruvate supplemented with 5 per cent fetal bovine serum (FBS) and divided into following four groups:

Gp.1 Control cells (untreated cells)

Gp.2 Irradiated cells (cells treated with gamma radiation)

Gp.3 Cells treated with radioprotective drug

Gp.4 Irradiated cells pretreated with radioprotective drug

3.8.1.1.2 When cells were reached approximately at 80 per cent confluency, the medium removed by suction and cells harvested by trypsinization.

3.8.1.1.3 Cells were then pelleted by centrifugation at 800xg in a swinging-bucket rotor centrifuge at room temperature and the pellet was washed twice in phosphate-buffered saline (PBS).

3.8.1.1.4 Cells were counted in a hemocytometer or in an automated cell counter.

3.8.1.2 Measurement of ATP Synthesis

3.8.1.2.1 Cells were resuspended at 1×10^7 cells/ml in buffer A (Table 1) at room temperature.

3.8.1.2.2 Once resuspended in buffer A, cells were kept at room temperature for up to 15 minutes without loss of ATP synthesis activity.

3.8.1.2.3 However, when measuring multiple samples, trypsinization of one cell line at a time is recommended.

3.8.1.2.4 160 µl of cell suspension was incubated for 1 minute at room temperature with 50µg/ml digitonin with gentle agitation.

3.8.1.2.5 Cells were washed by adding 1 ml of buffer A, and pelleted at 800xg.

3.8.1.2.6 The cell pellet was resuspended in 160 µl of buffer A and 0.15mM P^1, P^5-di (adenosine) pentaphosphate, either 1mM malate plus 1mM pyruvate or 5mM succinate plus 2 µg/ml rotenone, 10 µl of buffer B (containing 0.8 mM luciferin and 20 µg/ml luciferase in 0.5M Tris–acetate pH 7.75) and 0.1mM ADP were added.

3.8.1.2.7 For each sample, one replicate tube was prepared containing the above components plus 10 µg/ml oligomycin.

3.8.1.2.8 The ATPase inhibitor oligomycin was added to obtain the baseline luminescence corresponding to non-mitochondrial ATP production. (oligomycin has no effect on luciferase activity).

3.8.1.2.9 Cells were placed in a counting luminometer, and the light emission was recorded in kinetic mode.

3.8.1.2.10 The integration time for each reading was set at 1s and the interval between readings; 15s, for a total recording time of 5 minutes (20 readings).

3.8.1.2.11 In control cells, luminescence increased linearly for approximately 2.5 minutes. It reached a peak at 4–5 minutes, followed by a progressive decrease. The decrease in luminescence was probably due to the chemical properties of luciferin, which is progressively transformed into the inactive derivative deoxyluciferin.

3.8.1.2.12 The linear portion of the curve was used to extrapolate the variation in luminescence per unit of time (D relative light units or DRLU).

3.8.1.2.13 The DRLU measured in the presence of oligomycin was subtracted from the total DRLU.

3.8.1.2.14 The D luminescence was then converted into ATP concentration based on the values obtained from an ATP standard curve (luminescence obtained with luciferase was directly proportional to ATP concentration in the physiological concentration range.

3.8.1.2.15 A standard ATP/luminescence curve was constructed by measuring luminescence derived from ATP solutions containing 0, 0.05, 0.1, 0.25, 0.5, and 1.0mM ATP in buffer A and 10 µl of buffer B (1-7).

Table 1: Compositions and Storage Conditions of Reagents and Buffers Used in the Assay

Sl.No.	Compound	Volume (Final Concentration)	Stock Solution	Remark
1.	Buffer A (cell suspension)	160 ml	150 mMKCl, 25 mMTris–HCl, 2mM EDTA, 0.1 per cent bovine serum albumin, 10 mM potassium phosphate, 0.1mM MgCl, pH 7.4	Store at 4°C; at room temperature
2.	Digitonin	5 µl (50 µg/ml)	2 mg/ml in buffer A	Prepare the same day; keep on ice
3.	P^1, P^5-Di (adenosine) pentaphosphate	5 µl (0.15mM)	6mM in water	Store at –20°C; thaw and keep on ice
4.	Malate^a	2.5 µl (1mM)	80mM in buffer A	1M stock in water; store at –20 °C; thaw and dilute
5.	Pyruvate$^\$$	2.5 µl (1mM)	80mM in buffer A	1M stock in water; store at –20°C; thaw and dilute
6.	Succinate^a	5 µl (5mM)	200mM in buffer A	1M stock in water; store at –20°C; thaw and dilute
7.	Rotenone	2 µg/ml	400 µg/ml in DMS	Store at –20°C; keep on ice
8.	Oligomycin	2 µl (2µg/ml)	0.2 mg/ml in ethanol	Store at –20 °C; keep on ice
9.	ADP	5 µl (0.1mM)	4mM in buffer A	Prepare the same day; keep on ice
10.	Luciferin	See buffer B	100mM in water	Store at –20°C; thaw and keep on ice
11.	Luciferase	See buffer B	1 mg/ml in 0.5M Tris–acetate, pH 7.75	Store at –20°C; thaw and keep on ice
12.	Buffer B	10 µl	0.5M Tris–acetate, pH 7.75, 0.8mM luciferin, 20 µg/ml luciferase	Prepare the same day; keep at room temperature (25 °C)

$^\$$ Malate+pyruvate or succinate cab be used alternatively as substrate

References

1.   Zhao J, Lin S, Huang Y, Zhao J, and Chen P R (2013). Mechanism-based design of a photoactivatable firefly luciferase. *J. Am. Chem. Soc.* 135 (20), 7410–7413.

2.   Liu H, Jiang Y, Luo Y and Jiang W (2006). A simple and rapid determination of ATP, ADP and AMP concentrations in pericarp tissue of Litchi fruit by high performance liquid chromatography. *Food Technol. Biotechnol.* 44 (4) 531–534.

3. Lee M S, Park W S, Kim Y H, Ahn W G, Kwon S H and Her S (2012). Intracellular ATP assay of live cells using PTD-conjugated luciferase. *Sensors*, 12, 15628-15637, doi:10.3390/s121115628.

4. Ando T, Imamura H, Suzuki R, Aizaki H, Watanabe T, Wakita T and Suzuki T (2012). Visualization and measurement of ATP levels in living cells replicating hepatitis C virus genome RNA. *PLOS Pathog.* 8, e1002561.

5. Finger S, Wiegand C, Buschmann H J and Hipler U C (2012). Antimicrobial properties of cyclodextrinantiseptics-complexes determined by microplate laser nephelometry and ATP bioluminescence assay. *Int.J. Pharm.* 436, 851–856.

6. Koop A and Cobbold P H (1993). Continuous bioluminescent monitoring of cytoplasmic ATP in single isolated rat hepatocytes during metabolic poisoning. *Biochem. J.* 295, 165–170.

7. Askgaard D S, Gottschau A, Knudsen K and Bennedsen J (1995). Firefly luciferase assay of adenosine triphosphate as a tool of quantitation of the viability of BCG vaccines. *Biologicals*, 23, 55–60.

Protocol No. 3.9: Measurement of Creatine and Phosphorylated Nucleotides Homeostasis in Irradiated Animal Tissue/Irradiated Cultured Cells and Role of Radioprotective Drug to Maintain it using HPLC Method

Assay Requirement

Six-well dishes or culture flasks, 1.5 ml microfuge tube, perchloric acid, refrigerated centrifuge, K_2CO_3 Para-film, 4.6-mm i.d.X150-mm, 3-μm particle-size C-18 HPLC column, NaH_2PO_4 tetra-butyl-ammonium, acetonitrile, rainin, Nylon-66 filter, creatine, HPLC system with UV-visible detector, gamma irradiator.

3.9.1 Assay Procedure

3.9.1.1 Experimental groups preparations

Animals or cells were divided into following four groups:

Gp.1 Control animals/cells (untreated animals/cells)

Gp.2 Irradiated animals/cells (animals/cells treated with gamma radiation)

Gp.3 Animals/cells treated with radioprotective drug

Gp.4 Irradiated animals/cells pretreated with radioprotective drug

Followed by drug and radiation treatments, samples were prepared

3.9.1.2 Sample Preparation

3.9.1.2.1 Mice were anesthetized and the whole animal was immediately frozen in liquid nitrogen.

3.9.1.2.2 Striatum, cortex, and cerebellum were quickly dissected on a cold plate at −20°C. Frozen tissues were transferred to a 1.5-ml microfuge tube and 10 μl of ice-cold 0.4M perchloric acid was added per milligram of tissue.

3.9.1.2.3 The tissue was immediately homogenized with a pellet pestle.

3.9.1.2.4 The acidic homogenate was kept on ice for 30 minutes and then centrifuged at 14,000rpm at 4°C for 10 minutes.

3.9.1.2.5 An aliquot of the pellets was set aside for protein measurements.

3.9.1.2.6 The supernatant was neutralized with 10μl of 4M K_2CO_3 added to 100μl of the supernatant, kept on ice for 10 minutes and at -80°C for 1-2h to promote precipitation of the perchlorate and then centrifuged again.

3.9.1.2.7 Supernatants were stored at -80°C until HPLC assay performed.

or

3.9.1.2.8 Cells were grown as described above in six-well dishes/culture flasks.

3.9.1.2.9 Culture medium was removed by aspiration followed by immediate addition of ice-cold 0.4M perchloric acid (500 μl per 1,000,000 cells).

3.9.1.2.10 The culture dish was sealed tightly with Para-film and cooled at -80°C.

3.9.1.2.11 Cell lysates were thawed on ice, scraped off the wells thoroughly and transferred to 1.5 ml microfuge tubes.

3.9.1.2.12 Samples were centrifuged at 14,000 rpm at 4°C for 10 minutes.

3.9.1.2.13 External standards were prepared in 0.4M perchloric acid, neutralized and treated in exactly the same way as the samples.

3.9.1.2.14 Protein measurements were performed using standard protocol.

3.9.1.3 Chromatography Conditions

3.9.1.3.1 The gradient elution was performed on a 4.6-mm i.d.X150-mm, 3-μm particle-size YMC C-18 HPLC column with two buffers at a rate of 1 ml per minute. Buffer A contained 25mM NaH_2PO_4, 100 mg/liter tetra-butylammonium (pH 5). Organic buffer B was composed of 10 per cent (v/v) acetonitrile in 200mM NaH_2PO_4, 100 mg/liter tetra-butyl-ammonium (pH 4.0).

3.9.1.3.2 Buffers were filtered through a Rainin 0.2 μm Nylon-66 filter and degassed in a flask linked with vacuum pipe.

3.9.1.3.3 The gradient was 100 per cent buffer A from 0– 5 minutes, 100 per cent buffer A to 100 per cent buffer B from 5–20 minutes, and 100 per cent buffer A from 20 to 31minutes for column re-equilibration, which was sufficient to achieve stable baseline conditions.

3.9.1.3.4 50μl of prepared sample or standard mixture was auto-injected and UV monitored at 210nm from 0 to 10 minutes for creatine detection and at 260nm from 10 to 31 minutes for phosphorylated nucleotides. Peaks were identified by their retention times and by using co-chromatography with standards.

3.9.1.4 Standard Curves Preparation

3.9.1.4.1 Each standard of interest was first subjected to chromatography individually to obtain its retention time and to be able to later identify each compound in a standard mixture.

3.9.1.4.2 A standard curve for each compound was constructed by plotting peak heights (μV) versus concentration (10–1000 μM for creatine and 5–500 μM for nucleotides).

3.9.1.4.3 Linear curves were plotted.

3.9.1.4.4 The quantification of creatine and nucleotides in the sample was carried out using the external standard calibration (*i.e.*, by co-chromatography of the mixed standard solution and samples) and integration of sample peak heights against corresponding standard curves (1-6).

References

1. Manfredi G, Yang L, Gajewski C D and Mattiazzi M (2002). Measurements of ATP in mammalian cells. *Methods*. 26, 317–326.

2. Xue X F, Wang F, Zhou J H, Chen F, LiY Y and Zhao J (2009). Online cleanup of accelerated solvent extractions for determination of adenosine 52-triphosphate (ATP), adenosine 52-diphosphate (ADP) and adenosine 52-monophosphate (AMP) in royal jelly using high-performance liquid chromatography. *J. Agric. Food Chem*. 57 (11), 4500-4505.

3. Liu H, Jiang Y, LuoY and Jiang W (2006). A simple and rapid determination of ATP, ADP and AMP concentrations in pericarp tissue of Litchi fruit by high performance liquid chromatography. *Food Technol. Biotechnol*. 44(4), 531–534.

4. Khlyntseva S V, Bazel Y R, Vishnikin A B and Andruch V (2009). Methods for the determination of adenosine triphosphate and other adenine nucleotides. *J. Analytical Chem*. 64 (7), 657-673.

5. Brugidou C, Rocher A, Giraud E, Lelong B, Marin B, and Raimbault M (1991). A new high performance liquid chromatographic technique for separation and determination of adenylic and nicotinamide nucleotides in *Lactobacillus plantarum. Biotechnol. Techniques*. 5 (6), 475-478.

6. Özogul F, Taylor K D A, Quantick P C and Özogul Y (2008). A rapid HPLC determination of ATP-related compounds and its application to herring stored under modified atmosphere. DOI: 10.1111/j.1365-2621.2000.00405.x.

Section 4

Determination of Radiation Induced Apoptosis and DNA Damage

Apoptosis is a genetically programmed mode of cell death that accompanied by numerous molecular, biochemical and morphological changes to the cellular architecture. Apoptosis not only regulate cell death but also insure the efficient removal of apoptotic bodies by phagocytes. Apoptotic cells generally characterized by a variety of common characteristics that include cell shrinkage, plasma membrane blebbing, cell detachment, nuclear condensation, DNA fragmentation, externalization of phosphatidylserine (PS) and activation of caspases. Whereas, necrotic cell death is characterized by organelle swelling and plasma membrane rupture with none of the features of apoptosis. Here, we highlight a number of assays, utilizing both microscopy and flow cytometry, to determine radioprotective efficacy of a drug molecule in terms of its potential to inhibit radiation induced apoptosis in irradiated cells/tissue.

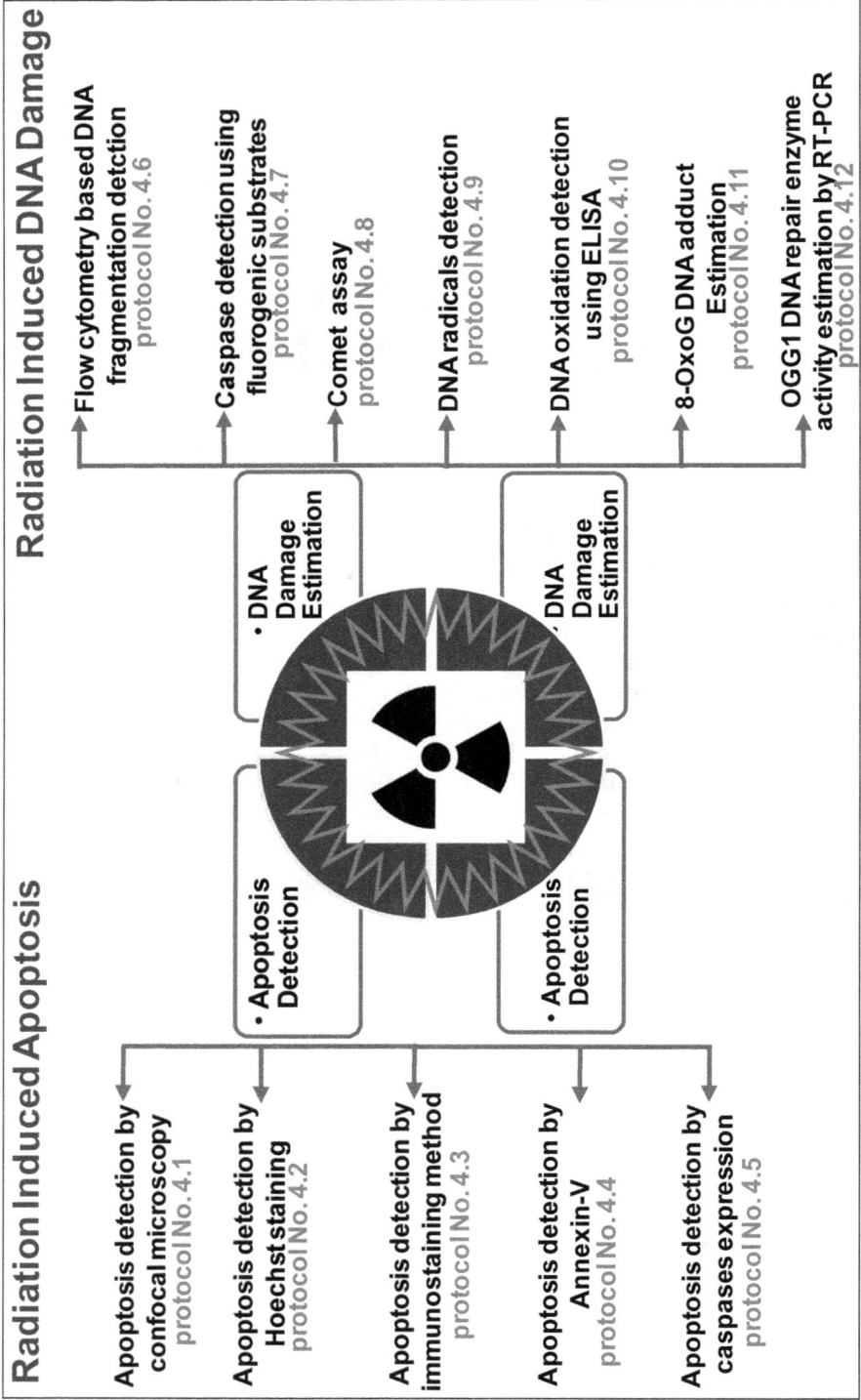

Figure 12: Schematic Summary of the Assays Used to Determine Radiation Induced Apoptosis and DNA Damage.

Protocol No 4.1: Studies on Morphological Features of Apoptosis Induced by Gamma Irradiation and Effect of Radioprotective Compounds on Apoptosis Inhibition using Phase Contrast and Confocal Microscopic Techniques

Assay Requirement

TRAIL ligand, PBS (pH 7.4), methanol, eosin dye, methylene blue dye, DPX mounting medium, culture dishes, culture medium, centrifuge, glass slides, CO_2 incubator, Biosafety cabinet level-II, inverted phase contrast/confocal microscope

4.1.1 Assay Procedure

4.1.1.1 Cells were cultured using standard procedures and divided into following four groups:

  i. Control cells (untreated)

  ii. Cells irradiated with gamma radiation (2-4 Gy)

  iii. Cells treated with radioprotective drug

  iv. Irradiated cells pretreated with radioprotective drug

4.1.1.2 Apoptosis in the cells (2×10^5 cells) was triggered by either incubation the cells with 100 ng/ml TRAIL ligand or by irradiation with gamma radiation.

4.1.1.3 After 12-24h, cells were harvested along with supernatant (which may contain detached or floated dead cells).

4.1.1.4 Cells were centrifuged at 400xg for 3 minutes. The cell pellet was re-suspended in PBS (pH 7.4) to achieve a cell number of 2×10^5/ml.

4.1.1.5 200µl of cell suspension was centrifuged at 250 rpm for 3 minutes.

4.1.1.6 Cells were now immobilised on slides and fixed with 100 per cent methanol. Dip the slides in methanol 10 times in quick succession.

4.1.1.7 Slides were then stained with eosin as outlined for methanol in Step 4.1.1.6. Allow excess eosin stain to drain from slide.

4.1.1.8 Transfer the slides to methylene blue stain. Dip slides in methylene blue 5 times in quick succession.

4.1.1.9 Slides were then re-rinsed in water to remove excess dye.

4.1.1.10 A drop of DPX mounting medium placed on a cover slip, which was then used to mount cells. Slides are now ready to examine. Condensation and fragmentation of nuclei that occurs during apoptosis was easily detected under the light microscope (1-11).

References

1. Aldridge D R, Arends M J and Radford I R (1995). Increasing the susceptibility of the rat 208F fibroblast cell line to radiation-induced apoptosis does not alter its clonogenic survival dose-response. *Br. J. Cancer* 71, 571–577. doi:10.1038/bjc.1995.111.

2. Hernández-Flores G, Gómez-Contreras P C, Domínguez- Rodríguez J R, Lerma-Díaz J M, Ortíz-Lazareno P C, Cervantes-Munguía R, Flores N J E S, Orbach-Arbouys S, Scott-Algara D and Bravo-Cuellar A (2005). α-irradiation induced apoptosis in peritoneal macrophages by oxidative stress: Implications of antioxidants in caspase mitochondrial pathway. *Anticancer Res.* 25: 4091-4100.

3. Mishra K P (2004). Cell membrane oxidative damage induced by gamma-radiation and apoptotic sensitivity. *J. Environ. Pathol. Toxicol. Oncol.* 23, 61-66.

4. Opferman J T and Korsmeyer S J (2003). Apoptosis in the development and maintenance of the immune system. *Nat. Immun* 4, 410-415, 2003.

5. Joya C, Afshin S and Sten O (2000). Trigggering and modulation of apoptosis by oxidative stress. *Free Rad. Biol. Med.* 29, 323-333.

6. Arends M J, Mcgregor A H and Wyllie A 1 1 (1994). Apoptosis is inversely relatei to necrosis and determine net growth in tumours bearing constitutively expressed myc, ras and HPV oncognes. *Am. J. Padul.* 144, 1045-1057.

7. Evan GL Wyllie Al, Gilbert CS, Lntlewood TD, Land H, Brooks M, Walters CM, Penn lz and Hancock DC (1992). Induction of apoptosis in fibroblasts by c-myc protein. *Cell.* 69, 119-128.

8. Tomei L D, Kanter P and Wenner C E (1988). Inhibition of radiation-induced apoptosis *in vitro* by tumor promoters. *Biochem. Biophys. Res. Commun.* 155, 324-331.

9. Ormerod M G, Sun X M, Brown D, Snowden R T and Cohen G M (1993). Quantification of apoptosis and necrosis by flow cytometry. *Acta Oncol.* 3: 417-423.

10. Cohen G M, Sun X M, Snowden R T, Dinsdale D and Skilleter D N. (1992). Key morphological features of apoptosis may occur in the absence of internucleosomal DNA fragmentation. *Biochm. J.* 236, 331-334.

11. Schmitz A, Bayer J, Dechamps N and Thomas G (2003). Intrinsic susceptibility to radiation-induced apoptosis of human lymphocyte subpopulations. *Int. J. Radiat. Oncol. Biol. Phys.* 57, 769-778

Protocol No 4.2: Detection of Apoptosis in Irradiated and Radioprotective Drug Treated Cells using Hoechst Staining

Assay Requirement

Cell culture dishes, Culture medium, Paraformaldehyde, PBS (pH 7.2), Triton X-100, Hoechst 33342, CO_2 incubator, Biosafety cabinet level-II, Fluorescence microscope, gamma irradiator.

4.2.1 Assay Procedure

4.2.1.1 Cells were cultured using standard procedures and divided into following four groups:

    i. Control cells (untreated)

    ii. Cells irradiated with gamma radiation (2-4 Gy)

    iii. Cells treated with radioprotective drug

    iv. Irradiated cells pretreated with radioprotective drug

4.2.1.2 After completion of incubation time followed by treatments, cells were fixed in 3.7 per cent paraformaldehyde for 10 minutes and slides prepared.

4.2.1.3 Slides were then washed with PBS (pH 7.2) followed by permeabilization of the cells using 0.15 per cent Triton X-100 in PBS (pH 7.2) for 15 minutes.

4.2.1.4 Washing step repeated and cells were incubated with Hoechst 33342 stain (500 nM) for 10 minutes.

4.2.1.5 Stained cells were washed with PBS (pH 7.2) before mounting with anti-fade medium and cover slips.

4.2.1.6 Slides were observed under the fluorescence microscope. Fluorescence dense spots were indicated the fragmented DNA and apoptotic bodies (1-6).

References

1. Belloc F, Dumain P, Boisseau M R, Jalloustre C, Reiffers J, BernardP and Lacombe F (1994). A flow cytometric method using Hoechst 33342 and propidium iodide for simultaneous cell cycle analysis and apoptosis determination in unfixed cells. *Cytometry* 17, 59-65.

2. Ellwart J Wand Dormer P (1990). Viability measurement using spectrum shift in Hoechst 33342 stained cells. *Cytometry* 11, 239-243.

3.   Lalande M E, Ling V, Miller R G (1981). Hoechst 33342 uptake dye as a probe of membrane permeability changes in mammalian cells. *Proc. Natl. Acad. Sci.* 78, 363.

4.   Loken MR (1980). Separation of viable T and B lymphocytes using a cytochemical stain, H33342. *J Histochem. Cytochem.* 28, 36-41.

5.   Zhang G, Gurtu V, Kain S R and Yan G (1997). Early detection of apoptosis using a fluorescent conjugate of Annexin V. *BioTechniques.* 23, 525-531.

6.   Allen S, Sotos J, Sylte M J, and Czuprynski C J (2001). Use of Hoechst 33342 Staining to detect apoptotic changesin bovine mononuclear phagocytes infected with *Mycobacterium avium* subsp. *Paratuberculosis. Clinical Diagnostic Laboratory Immunol.* 8(2), 460–464.

Protocol No. 4.3: Determination of Apoptosis in the Irradiated and Radioprotective Drug Treated Cells using Immunostaining Method

Assay Requirement

Cisplatin, paraformaldehyde, PBS (pH 7.2), Triton-X100, BSA, anti-TOM20 monoclonal antibodies, appropriate Alexa-488 or rhodamine conjugated secondary antibodies, Hoechst 33342, anti-fade medium, confocal microscope, gamma irradiator.

4.3.1 Assay Procedure

4.3.1.1 Apoptosis can be induced in the cells grown on the cover slips either by addition of the chemotherapeutic drug like cisplatin (50 µM) or gamma radiation treatment.

4.3.1.2 Cells were then washed with PBS (pH 7.2) and fixed with 3.7 per cent paraformaldehyde in PBS for 10 minutes.

4.3.1.3 Washing steps were repeated followed by permeabilization with0.15 per cent Triton-X100 in PBS for 15 minutes, and blocking in PBS with 2 per cent BSA for 30 minutes.

4.3.1.4 Mitochondria were stained with an anti-TOM20 monoclonal antibodies diluted 1:200 in PBS 2 per cent BSA for 1h at room temperature.

4.3.1.5 Slides were then washed and incubated with an appropriate secondary antibody (Alexa-488 or Rhodamine conjugated) diluted 1:1000 in PBS with 2 per cent BSA for 1h at room temperature.

4.3.1.6 Cells were washed as mentioned above and incubated with Hoechst 33342 (500 nM) for 10 minutes.

4.3.1.7 Slides were mounted with an anti-fade medium and observed on a laser scanning confocal microscope using a 488nm Argon laser (green fluorescence), a 543 nm He-Ne laser (red fluorescence) and a 405nm LD laser (1-4).

References

1. Schmid I, Uittenbogaart C H and Giorgi J V (1994). Sensitive method for measuring apoptosis and cell surface phenotype in human thymocytesby flow cytometry. *Cytometry* 15, 12-17.

2.   Ellwart J W and Dormer P (1990). Viability measurement using spectrum shift in Hoechst 33342 stained cells. *Cytometry*, 11, 239-243.

3.   Lalande ME, Ling V, Miller RG (1981). Hoechst 33342 uptake dye as a probe of membrane permeability changes in mammalian cells. *Proc. Natl. Acad. Sci.* 78, 363.

4.   Loken MR (1980). Separation of viable T and B lymphocytes using a cytochemical stain, H33342. *J Histochem. Cytochem*, 28, 36-41.

Protocol No. 4.4: Detection of Fragmented DNA in Irradiated Cells and its Protection by Radioprotective Drug using Flow Cytometry

Background

A feature that clearly distinguishes apoptosis from necrosis is DNA fragmentation and formation of high molecular weight (>50 kbp) and nucleosome-sized (200 bp) DNA fragments in apoptotic cells. Though, several biochemical techniques such as agarose gel electrophoresis is routinely used for detection of DNA ladders. However, these techniques are time consuming and do not allow individual cell analysis. Alternatively, a quantitative way to analyze DNA fragmentation is flow-cytometric detection of DNA hypoploidy after adding PI to the dying cells. PI intercalates with DNA is fragments and the size of DNA fragments appears as a hypoploid DNA histogram can be quantified (1-9).

Assay Requirement

Cell culture dishes, culture medium, anti-Fas antibody, phosphate-citrate wash buffer, ethanol, RNase A, trypsin, ethanol, propidium iodide, flow cytometer.

4.4.1 Assay Procedure

4.4.1.1 Cells were cultured using standard procedures and divided into following four experimental groups:

    i. Control cells (untreated)

    ii. Cells irradiated with gamma radiation (2-4 Gy)

    iii. Cells treated with radioprotective drug

    iv. Irradiated cells pretreated with radioprotective compound

4.4.1.2 In positive control groups, apoptosis can be induced in the cells by incubation with 200 ng/ml anti-Fas antibody for 1-4h.

4.4.1.3 Irradiated and drug treated cells were harvested by trypsinization and followed by centrifugation at 400xg for 5 minutes.

4.4.1.4 Cells were then washed with PBS (pH 7.2) and re-centrifuged at 400xg for 1 minute.

4.4.1.5 Cells were re-suspended in 1ml ice-cold 70 per cent ethanol and incubated for at least 1h at -20 °C for fixation (cells can be stored for upto 6 months at -20°C).

4.4.1.6 Followed by fixation, cells were centrifuged at 2500xg for 5 minutes.

4.4.1.7 Followed by centrifugation, aspirate off the ethanol without disturbing the cell pellet and re-suspend them with 1 ml phosphate-citrate wash buffer (composition: 200 mM Na_2HPO_4, 100 mM citric acid) followed by centrifugation at 2500xg for 1 minutes.

4.4.1.8 To stain the nuclei, cells were incubated for 30 minutes with PBS (pH 7.2) containing propidium iodide 10 μg/ml and RNase A 100 μg/ml.

4.4.1.9 Now stained cells were subjected to analysis of fragmented DNA using flow cytometer (1-8).

References

1. Facchinetti A, Tessarollo L, Mazzocchi M, Kingston R, Collavo D and Biasi G (1991). An improved method for the detection of DNA fragmentation. *J. Immunol. Methods.* 136: 1251-1256.

2. Wijsman J H, Jonker R R, Keijzer R, Van de Velde C J H, Cornelisse C J and Van Dierendonck J H (1993). A new method to detectapoptosis in paraffin sections: in situ end-labeling of fragmented DNA. *J. Histochem. Cytochem.* 41: 7-12.

3. Nicoletti I, Migliorati G, Pagliacci M C, Grignani F, Riccardi C (1991). A rapid and simple method for measuring thymocytes apoptosis by propidium iodide staining and flow cytometry. *J. Immunol. Methods.* 139: 271-279.

4. Telford W G, King L E and Fraker P J (1992). Comparative evaluation of several DNA binding dyes in the detection of apoptosis-associated chromatin degradation by flow cytometry. *Cytometry.* 13: 137-143.

5. Afanasev VN, Korol BA, Mantsygin YA, Nelipovich PA,PechatnikovV A and Umansky S R (1986). Flow cytometryand DNA degradation of two types of cell death. *FEDS Lett.* 194: 347-350.

6. Brown DG, Sun XM and Cohen GM. (1993). Dxamethasone induced apoptosis involves cleavage of DNA to lare fragments pron to inter nucleosomal fragmentation. *J. Biow. Chem.* 26: 3037-3039.

7. Cohen GM, Sun XM, Snowden RT, Dinsdale D and Skilleter DN(1992). Key morphological features of apoptosis may occur in the absence of inter nucleosomal DNA fragmentation. *Biochem. J.* 236: 331-334.

8. Radford IR and Murphy TK (1994). Radiation response of mouse lymphoid and myeloid cell lines. III. Different signals can lead to apoptosis and may influence sensitivity to killing by DNA double strand breakage. *Int. J. Radiat. Biol.* 65: 229-239.

Protocol No. 4.5: Determination of Radiation Induced Cell Death and Role of Radioprotective Drug to Reverse Cell Death using FITC-Conjugated Annexin-V

Background

Annexin-V-FITC fluorescence (525 nm) can be measured with green channel and PI fluorescence (617 nm) can be measured with red channel. Samples can be analyzed by a single parameter histogram (FL1) or two-parameter dot-plots (FL1 vs. FL2). For dot plot analysis, cells in the early stages of apoptosis will be located on the bottom right quadrant of the dot-plot as single positive Annexin V binding cells, since at this stage cell membranes are still intact and PI cannot enter the cells. In later stages of apoptosis as the membrane becomes more permeable, PI can enter cells leading to a double positive population. In contrast cells undergoing necrotic cell death go directly to the double positive state.

Assay Requirement

Cell culture dishes, culture medium, Annexin V-FITC, Annexin-V-binding buffer with HEPES-NaOH, Ca^{2+} ions, NaCl, KCl, $MgCl_2$, $CaCl_2$, propidium iodide, flow cytometer.

4.5.1 Assay Procedure

4.5.1.1 Cells were cultured using standard procedures and divided into following four treatment groups:

    i. Control cells (untreated)

    ii. Cells irradiated with gamma radiation (2-4 Gy or more)

    iii. Cells treated with radioprotective drug

    iv. Irradiated cells pretreated with radioprotective drug

4.5.1.2 Irrespective of the cell line being used, the assay is based on the 15 minutes incubation of the cells in a solution containing Ca^{2+} ions and Annexin V-FITC (final concentration of $1\mu g/ml$).

4.5.1.3 The annexin-V-binding buffer consists of 10 mM HEPES-NaOH (pH 7.4), 150 mM NaCl, 5 mM KCl, 1 mM $MgCl_2$ and 1.8 mM $CaCl_2$. This-buffer should be stored at 4°C.

4.5.1.4 Followed by incubation propidium iodide can be mixed to facilitate the identification of cells undergoing secondary necrosis.

4.5.1.5 Propidium iodide was added to the cells suspension at a final concentration of 10µg/ml.

4.5.1.6 Analysis of annexin-V/PI-stained cells by flow cytometry allows for the quantitation of the cells that are (i) Annexin V-negative and PI-negative (live cells), (ii) Annexin V-positive and PI-negative (cells should be in early apoptotic phase), and (iii) Annexin V-positive and PI-positive (cells should be in late apoptotic and necrotic phase) (1-7).

Precautions

A number of considerations are important to ensure a successful assay:

1. The presence of Ca^{2+} ions is vital for annexin-V to bind phosphatidyl serine efficiently. The ideal concentration of Ca^{2+} ions *i.e.* 1.8 mM is recommended. Higher concentrations can lead to non-specific binding to other phospholipids such as phosphatidylcholine.

2. Propidium iodide is cytotoxic to the cells. To limit the toxicity, add PI at 10µg/ml (final concentration) and avoid prolong incubation before analysis. Another reason to avoid prolonged incubation time with PI is that some time viable cells will also eventually take up the dye may lead to false positive results.

3. Annexin V-FITC optimum concentrations was standardized to be between 0.5–1 µg/ml, however a titration of the stock is highly recommended to establish the optimal concentration for each protein batch (1-4).

References

1. Zhang G, Gurtu V, Kain SR and Yan G (1997). Early detection of apoptosis using a fluorescent conjugate of Annexin V. *BioTechniques*. 23, 525-531.

2. Koopman G, Reutelingsperger C P M, Kuijten G A M, Keehnen R M J, Pals S T and Van Oers M H J (1994). Annexin V for flow cytometric detection of phosphatidylserine expression on B cells undergoing apoptosis. *Blood*. 84, 1415-1420.

3. Homburg CHE, De-Haas M, Von Dem Borne A EG K, Verhoeven A J, Reutelingsperger C P M and Roos D (1995). Human neutrophils lose their surface FcRIII and acquire annexin V binding sitesduring apoptosis *in vitro*. *Blood*. 85, 532-540.

4. Vermes I, Haanen C H, Nakken S and Reutelingsperger C (1995). A novel assay for apoptosis flow cytometric detection of phosphatidylserine expression on early apoptotic cells using fluorescein labelled annexin V. *J. Immunol. Methods* 184, 39-51.

5. Van Engeland M, Nieland L J, Ramaekers FC, Schutte B, Reutelingsperger C P (1998). Annexin V-affinity assay: a review on an apoptosis detection system based on phophatidylserine exposure. *Cytometry*. 31, 1-9.

6. http://www.kumc.edu/Documents/flow/Annexin per cent 20V_PI.pdf.

7. http://www.immunochemistry.com/products/annexin-v-fitc-apoptosis-detection-kit-for-flow-cytometry.html.

Protocol No. 4.6: Estimation of Caspases Mediated Cell Death using FLICA in Irradiated Cells and its Inhibition with Radioprotective Drug Pretreatment

Background

Gamma radiation induced cell death via capases activation. Therefore, inhibition of apoptosis by radioprotective drug can provide survival benefits to the irradiated cells. In view of that, FAMFLICA-fmk can be used as capases inhibitor in the assay in positive control group. FAMFLICA-fmk acts as caspase inhibitor, and if added to the cells prior (similarly as radioprotective drug prior treatment) to irradiation, it would inhibit subsequent caspase activation and apoptosis. Note that caspase activation time may vary with different cell lines.

Assay Requirement

Cell culture dishes, culture medium, FAM-FLICA, DMSO, PBS, propidium Iodide, flow cytometer, centrifuge and gamma irradiator.

4.6.1 Assay Procedure

4.6.1.1 In brief, cells were cultured using standard procedures and divided into following seven groups:

☆ Control cells (untreated)

☆ Cells irradiated with gamma radiation (2-4 Gy or desired dose)

☆ Cells treated with radioprotective drug

☆ Irradiation cells pretreated with radioprotective drug.

☆ Cells treated with FAM-FLICA

☆ Irradiated cells pretreated with FAM-FLICA

☆ Irradiated cells pretreated with radioprotective drug followed by FAM-FLICA treatment

4.6.1.2 Cells (2×10^5 cells/ml) were triggered to undergo apoptosis by gamma radiation exposure.

4.6.1.3 FAM-FLICA was reconstituted with 50 µl DMSO to give a stock solution (150x).

4.6.1.4 This stock solution can be stored for upto 6 months at -20 °C. Prior to addition of FAM-FLICA to the cells, dilute the stock 1:5 with 200µl PBS (pH 7.2) to achieve a working concentration of 30X.

4.6.1.5 At the appropriate time points cells were centrifuged at 400xg for 5 minutes and resuspend again in 290µl culture medium. Add 10µl of the 30x FAM-FLICA stock to control and irradiated/drug treated cells.

4.6.1.6 Cells were incubated for 30 minutes at 37 °C to allow the binding of FAM-FLICA to activated caspases.

4.6.1.7 To ensure the efficient FLICA staining, gently resuspend the cells by swirling them every 15–20 minutes.

4.6.1.8 1.0 ml PBS (wash buffer) added to each sample and cells were harvested by gentle centrifugation at 400xg for 5 minutes.

4.6.1.9 Resuspend the cells again in 1.0 ml wash buffer. Incubate for 10 minutes to allow any unbound FAM-FLICA to diffuse out from the cells. Repeat above centrifugation step once more.

4.6.1.10 Resuspend the cells in 200µl wash buffer containing propidium Iodide at 10µg/ml concentration.

4.6.1.11 Cells are now ready for analysis by flow cytometer (1-6).

References

1. Henry C M, Hollville E, Martin S J (2013). Measuring apoptosis by microscopy and flow cytometry. *Methods.* 61, 90–97.

2. Ravichandran K S (2011). Beginnings of a good apoptotic meal: the find-me and eat-me signaling pathways. *Immunity.* 35, 445–455.

3. Martin S J, Henry C M, Cullen S P (2012). A perspective on mammalian caspases as positive and negative regulators of inflammation. *Mol. Cell.* 46, 387-397.

4. Slee E A, Harte M T, Kluck R M, Wolf B B, Casiano C A, Newmeyer D D, Wang H G, Reed J C, Nicholson D W, Alnemri E S, Green D R, Martin S J (1999). Ordering the cytochrome cinitiated caspase cascade: Hierarchical activation of caspases-2,-3,-6,-7,-8, and -10 in a caspase-9–dependent manner. *J. Cell Biol.* 25, 281–292.

5. Coleman M L, Sahai E A, Yeo M, Bosch M, Dewar A, Olsen M F (2001). Membrane blebbing during apoptosis results from caspase-mediated activation of ROCK I. Nat. *Cell Biol.* 4: 339–345.

6. Enari M, Sakahira H, Yokoyama H, Okawa K, Iwamatsu A, Nagata S (1998). A caspase-activated DNase that degrades DNA during apoptosis and its inhibitor ICAD. *Nature.* 391, 43-50.

Protocol No. 4.7: Analysis of Gamma Radiation Induced Caspase Activation and its Modulation by Radioprotective Drug using Fluorogenic Tetrapeptide Substrates

Box Material

To measure the activity of caspases instead of their proteolytic activation state, we can use fluorogenic tetrapeptide substrates. These tetrapeptide P4–P1 substrates fit in the catalytic pocket of caspases and are recognized by the corresponding S4–S1 residues of the protease. The cleavage occurs after the aspartic acid residue at P1 is coupled to 7-amino-4 methylcoumarin (AMC) or 7-amino-4-trifluoromethylcoumarin (AFC). Hydrolysis of this peptide bond releases free AMC or AFC and the emitted fluorescence can be measured by a fluorospectrometer. In this way acetyl (Ac)-DEVD-AMC was used as a preferred substrate for caspase-3, -7, Ac-LEHD-AMC for caspase-5, Ac-YVAD-AMC for caspase-1 and Ac-IETD-AMC for caspase-8 and 6 and Ac-WEHD-AMC for caspase-1, -4 and 5. However, the claimed specificities of the peptide substrates should be viewed with caution. The problem with this assay is that peptide substrate specificity is often abolished by high caspase concentrations. Cellular concentrations of caspase-3 also cleave Ac-IETD-AMC (substrate for caspase-8 and -6) even more efficiently than caspase-8 and 6, due to the high Kcat value of caspase-3. Therefore, enzymatic measurements of caspases activities must be combined with immunoblotting to identify the presence and activation status of the caspases (1-4).

Assay Requirement

Eppendorf tubes, PBS, lysis buffer, Tris-HCl, NaCl, $MgCl_2$ PMS, aprotinin, leupeptin, oxidized glutathione, Bradford reagent, dithiothreitol, GSH, CSF buffer, fluorogenicsubstrate *i.e.* Ac-DEVD-AMC, Ac LEHD-AMC, Ac-YVAD-AMC, Ac-IETD-AMC and Ac- WEHD-AMC, gamma irradiator, fluorescence-spectrophotometer.

4.7.1 Assay Procedure

4.7.1.1 Approximately, 6×10^5 cells per well were seeded in six-well tissue culture plates.

4.7.1.2 Next day, cells were harvested and treated with gamma radiation in presence or absence of radioprotective drug in 1.5-ml Eppendorf tubes. Cells were then transferred to culture dishes and incubated at 37°C with 5 per cent CO_2 for 6-24h.

4.7.1.3 Cells were harvested and then washed with PBS by centrifugation for 5 minutes at 250xg.

4.7.1.4 The cells were lysed in 150µl cold lysis buffers (1 per cent NP-40, 10 mM Tris HCl (pH 7.4), 10 mM NaCl, 3 mM MgCl₂).

4.8.1.5 Protease inhibitors must be added to lysis buffer directly before cell lysis: (Protease inhibitors composition: 1mM PMSF, 0.3mM aprotinin, and 1mM leupeptin).

4.7.1.6 To block the catalytic cysteine of caspases and prevent the activation of caspase cascades, 1 mM of oxidized glutathione (GS-SG) was added during lysis at 4°C.

4.7.1.7 Cells debris was removed by centrifuging the lysate for 10 minutes at 20,800xg (4°C) and the cytosol was transferred to another Eppendorf tube (cytosolic cell lysate).

4.7.1.8 After protein concentration determination by Bradford method, the equivalent of 10µg protein was transferred to a new Eppendorf tube for caspase activity measurement, while the remainder was used for immunoblotting.

4.7.1.9 Importantly, before measurement of caspase activity in CSF buffer, 10mM dithiothreitol was added to remove the GSH from the catalytic cysteine.

4.8.1.10 Caspase activity was measured by incubating 10µg cytosolic cell lysate with 50µM of the fluorogenic substrate, *e.g.* Ac-DEVD-AMC in 150µl CFS-buffer (Composition: *10 mM HEPES NaOH, pH 7.4, 220 mM mannitol, 68 mM sucrose, 2 mM MgCl₂, 2 mM NaCl, 2.5 mM H₂KPO₄, 0.5 mM EGTA, 0.5 mM sodium pyruvate and 0.5 mM L-glutamine. PBS consists of 137 mM NaCl, 10 mM phosphate and 2.7 mM KCl, pH of 7.4)*

4.7.1.11 Dithiothreitol (DTT) was added to a final concentration of 10mM.

4.7.1.12 Caspase fluorogenic substrates (*i.e.* Ac- DEVD-AMC, Ac LEHD-AMC, Ac-YVAD-AMC, Ac-IETD-AMC, Ac- WEHD-AMC) were prepared as 100mM stock solutions in DMSO.

4.7.1.13 The release of fluorescent AMC was monitored at every two minutes for 1h using fluorescence spectrophotometer at 37 °C, using a filter with an excitation wavelength of 360nm and emission wavelength of 460nm (1-7).

4.7.1.14 The data can be expressed as the time-rate of increase in fluorescence (fluorescence units/min).

References

1. Berghe T V, Grootjans S, Goossens V, Dondelinger Y, Krysko D V, Takahashi N, and Vandenabeele P (2013). Determination of apoptotic and necrotic cell death *in vitro* and *in vivo*. *Methods*. 61, 117-129.

2. Denecker G, Vercammen D, Steemans M, Berghe T V, Brouckaert G, Loo G V, Zhivotovsky B, Fiers W, Grooten J, Declercq W and Vandenabeele P (2001). Death receptor-induced apoptotic and necrotic cell death: differential role of caspases and mitochondria *Cell Death Differ*. 8, 829–840.

3. Lamkan M, Declercq W, Depuydt B, Kalai M, Saelens X and VandenabeeleP (2003). The caspase family. In: M. Los, H. Walczak (Eds.), *Caspases: Their Role in Cell Death and Cell Survival*, TX: Landes Bioscience, Kluwer Academic Press, Georgetown.

4. Thornberry N A, Rano T A, Peterson E P, Rasper D M, Timkey T, Garcia-Calvo M, Houtzager V M, Nordstrom P A, Roy S, Vaillancourt J P, Chapman K T and Nicholson DW (1997). A combinatorial approach defines specificities of members of the caspase family and granzyme B. Functional relationships established for key mediators of apoptosis. *J. Biol. Chem.* 272, 17907–17911.

5. Talanian R V, Quinlan C, Trautz S, Hackett M C, Mankovich J A, Banach D, Ghayur T, Brady K D and Wong W W (1997). Substrate specificities of caspase family proteases. *J. Biol. Chem.* 272, 9677-9682.

6. Pop C and Salvesen G S (2009).Human caspases: activation, specificity, and regulation. *J. Biol. Chem.* 284 21777–21781.

7. Berghe T V, Kalai M, Loo G V, Declercq W and Vandenabeele P (2003).Fas induced necrosis. *J. Biol. Chem.* 278: 5622–5629.

Protocol No. 4.8 Screening of Compounds able to Inhibit Radiation Induced Single Strands Breaks in Genomic DNA in Cells using Comet Assay

Assay Requirement

Cell culture dishes, flasks, culture medium, cell culture facility, Ca^{2+}/Mg^{2+} free 1X PBS (pH 7.2), agarose, lysis buffer, NaOH, EDTA, ethanol, SYBR Green, fluorescence microscope with imaging system, comet score software and gamma irradiator.

4.8.1 Assay Procedure

4.9.1.1 Cells were grown using standard cell culture methods up to 90 per cent confluency.

4.8.1.2 Equal number (1000-2000) of cells was plated in 60mm culture discs in the following four experimental groups (sets) and incubated for 24h at 37°C, 5 per cent CO_2 and 95 per cent humidity:

  Set- A: Control cells (untreated cells).

  Set-B: Cells treated with different concentrations of compound.

  Set-C: Cells irradiated with gamma radiation (2-4 Gy) in absence of radioprotective compound.

  Set-D: Cells irradiated with gamma radiation (2-4 Gy) in presence of radioprotective compound.

4.8.1.3 Followed by different treatments, cells of all four groups were washed twice in ice-cold Ca^{2+}/Mg^{2+}free 1X PBS (pH 7.2).

4.8.1.4 Cells were mixed with 100µl of pre-warmed low melting agarose (1:10, v/v) and plated onto a slide (Comet slide).

4.8.1.5 After agarose solidified and attached to the slides, the slides were immersed in pre-chilled lysis solution [(Composition: 2.5 M NaCl, 100 mM EDTA, 10 mM Tris–HCl (pH 10.0), 1 per cent triton X100,10 per cent DMSO) for 1h on ice and then in alkaline unwinding solution (300 mM NaOH, 1 mM EDTA) for 1h at room temperature.

4.8.1.6 Electrophoresis was performed in pre-chilled alkaline electrophoresis solution (300mM NaOH, 1mM EDTA) at 4 °C for 45 minutes at 1 V/cm.

4.8.1.7 The slides were then washed twice with distilled H_2O for 5 minutes, followed by 5-minutes incubation with 70 per cent ethanol.

4.8.1.8 The slides were air-dried at 40°C for 15minutes in the dark and the agarose gels were stained with SYBR Green for 5 minutes at 4°C.

4.8.1.9 After SYBR Green was removed and slides were air-dried, images were acquired using a fluorescence microscope with a Cool SNAP digital camera and the imaging software. The levels of DNA damage were determined using Comet Score software (1-11).

References

1. Speit G and Hartmann A (2005). The comet assay: a sensitive genotoxicity test for the detection of DNA damage. *Methods Mol. Biol.* 291, 85-95.

2. Speit G and Hartmann A (2006). The comet assay: a sensitive genotoxicity test for the detection of DNA damage and repair. *Methods Mol. Biol.* 314, 275-286.

3. Speit G and Rothfuss A (2012). The comet assay: a sensitive genotoxicity test for the detection of DNA damage and repair. *Methods Mol. Biol.* 920, 79-90.

4. Zeni O and Scarfi M R (2010). DNA damage by carbon nanotubes using the single cell gel electrophoresis technique. *Methods Mol. Biol.* 625, 109-119.

5. Collins A R (2004). The comet assay for DNA damage and repair: principles, applications, and limitations. *Mol. Biotechnol.* 26(3), 249-261.

6. Shaposhnikov S A, Salenko V B, Brunborg G, Nygren J and Collins A R (2008). Single-cell gel electrophoresis (the comet assay): loops or fragments? *Electrophoresis.* 29(14), 3005-3012.

7. Olive P L and Banáth J P (2006). The comet assay: A method to measure DNA damagein individual cells. *Nature Protocols.* 1 (1), 23-29.

8. Ostling O and Johanson K J (1984). Microelectrophoretic study of radiation-induced DNA damages in individual mammalian cells. *Biochem. Biophys. Res. Commun.* 123, 291-298.

9. Olive P, Banát J P and Durand R E (1990). Heterogeneity in radiation-inducedDNA damage and repair in tumor and normal cells measured using the "comet" assay. *Radiat. Res.* 122, 86-94.

10. Banath J P, Kim A and Olive P L (2001). Overnight lysis improves the efficiency of detection of DNA damage in the alkaline comet assay. *Radiat. Res.* 155: 564–571.

11. Olive P L, Johnston P J, Banáth J P and Durand R E (1998). The comet assay: Anew method to examine heterogeneity associated with solid tumors. *Nat. Med.* 4, 103–105.

Protocol No. 4.9: Evaluation of Radiation Induced DNA Oxidation (DNA Radicals Detection) and its Protection by a Radioprotective Agent using Agarose Gel Electrophoresis Method

Assay Requirement

Nitrocellulose, AG501-X8 resin, DMPO, Immobilon-FL-poly-vinylidene difluoride (PVDF) membrane, LumiGLO peroxidase chemiluminescence substrate kit. Reacting-Bind DNA coating solution, goat anti-mouse IgG (H+L) conjugated to horse radish-peroxidase(HRP), rabbit anti-chicken IgY(H+L) conjugated to HRP, Donkey anti-chicken IgG-800 (H+L), IR Dye 800CW, donkey anti-mouseIgG-800(H+L), 10x orange loadingdye, Low IgG fetal bovine serum, Iscove's modified Dulbecco's medium, 6 per cent (w/v) DNA retardation poly-acrylamide gels, TBE running buffer, SYTO 60 red fluorescent nucleic acid stain, Calf thymus DNA, copper(II) chloride, casein, diethylenetri-amine-penta acetic acid (DTPA), polydeoxyguanylic acid, poly deoxycytidylic acid sodium salt (poly(dG), poly(dC), poly(deoxyguanylic–deoxycytidylic) acid sodium salt (poly(dG-dC), poly(dG-dC), polydeoxyadenylic acid, polythymidylic acid sodium salt (poly(dA)-(dT)), poly (deoxy-adenylic–thymidylic) acid sodium salt [poly-(dA-dT)], glyoxaltri-merdihydrate, chicken polyclonal antibody to anti-5,5- dimethyl-2-(8-octonoic acid)-1-pyrrolone-N-oxide conjugated to bovine serum albumin (anti-DMPO adduct) and mouse anti-DMPO adduct monoclonal antibody.

4.9.1 Preparation of DNA Radicals and Spin Trapping with DMPO

4.9.1.1 Following experimental groups were formed:

Set 1: Control DNA.

Set 2: DNA irradiated either with fix dose (>3kGy) or different doses in increasing order (*i.e.* 3, 5,7,10 kGy) of gamma radiation in absence of radioprotective compound in PBS.

Set 3: DNA irradiated either with fix dose (>3kGy) or different doses in increasing order (*i.e.* 3, 5,7,10 kGy) of gamma radiation in presence of radioprotective compound in PBS.

4.9.1.2 Followed by irradiation, copper (II) chloride in PBS and DMPO in PBS **(Composition: 2mM potassium phosphate, 8mM sodium phosphate, 2.7mM potassium chloride and 137mM sodium chloride, pH7.4)** was added in all sets of reaction mixtures.

4.9.1.3 Reaction mixture was then incubated for 30 minutes to 1h at 37°C temperature.

4.9.1.4 Followed by incubation, DTPA was added to a final concentration of 1mM to terminate the reaction.

4.9.1.5 The DNA was then precipitated with 1/10 volume of 3M sodium acetate, pH 5.2 and 2 volumes of ice-cold ethanol and incubated for 10 minutes at room temperature. (*Note: DMPO will be precipitated in this mixture if incubated at 4 °C*).

4.9.1.6 The DNA was centrifuged at 13,000rpm for15 minutes at room temperature, washed with 70 per cent (v/v) ethanol and re-dissolved in 10mM Tris, 1mM EDTA buffer (pH 8.0).

4.9.2 DNA Electrophoresis

4.9.2.1 Irradiated and controlled DNA was denatured by adding deionized formamide to a final concentration of 60 per cent (v/v), with 1/10 volume 10X orange loading dye and 1 µl 5µM SYTO 60 nucleic acid stain (for sample volumes ranging from 10 to 30µl).

4.9.2.2 The samples were than denatured by heating for 5 minutes at 65°C, followed by immediate chilling on ice for 5 minutes before loading on to the gel.

4.9.2.3 DNA to be run under native conditions, was mixed with 1/5 volume 10x orange loading dye and 1µl 5µM SYTO60 and incubated for 5 minutes at room temperature (Optional).

4.9.2.4 DNA (5µg/lane) was electrophoresed on either 1 per cent (w/v) agarose gels in TAE (40mMTris–acetate, 1mM EDTA) for 45 minutes at 90V.

or

on 6 per cent (w/v) DNA retardation polyacrylamide gels in 0.5xTBE (44.5mMTris, 44.5mM borate, 1 mM EDTA) for 90 min at 100V.

4.9.2.5 The DNA stained with SYTO 60 was visualized by scanning the gels using the 700nm channel on the infrared imaging system.

4.9.3 DNA Transfer to Membrane

4.9.3.1 Agarose gels were equilibrated in 20XSSC (3M sodium chloride, 300 mM sodium citrate) buffer with two 10-20 minutes incubations at room temperature (RT) with constant gentle agitation.

4.9.3.2 DNA was transferred to a nitrocellulose membrane by down ward capillary transfer overnight in 20X SSC buffer.

4.9.3.3 UV cross linker was used for UV (254nm) irradiation of the nitrocellulose membrane to crosslink the DNA using an exposure of 100-120,J/cm^2.

or

4.9.3.4 PVDF membranes were pre-wet in methanol for several seconds and then equilibrated in 0.5XTBE for 15 minutes.The DNA from polyacrylamide gels was then transferred on to the PVDF membrane by electro-blotting for 20 minutes at 15V on aTrans-Blot SD semi-dry transfer cell.

Note: Exposing PVDF membranes to UV crosslinking did not seem to improve binding of the DNA to the membrane, as the UV-cross-linked membranes gave the same DMPO signal as membranes that had not been subjected to UV crosslinking.

4.9.4 DNA Immunoblot

4.9.4.1 The Immobilon-FL poly-vinylidene difluoride (PVDF) membrane was blocked with 1 per cent (w/v) casein in PBS containing 5 µg ml^{-1} monoclonal anti-DMPO antibody or 10 µg ml^{-1} polyclonal anti-DMPO antibodies for 1h and washed the membrane in PBS for 5 minutes each three times.

4.9.4.2 Membrane was treated with IR-Dye-800CW donkey anti-mouse IgG (H+L) (1:15,000 dilution) or IR-Dye 800CW donkey anti-chicken IgG (H+L) (1:15,000 dilution) antibodies in blocking solution or in PBS.

(Note: As the chicken polyclonal anti-DMPO antibody bound nonspecifically to residual agarose on the membrane, the poly-clonal antibody was pre-incubated with 1 per cent (w/v) agarose in blocker for 2h at room temperature, and then the mixture was centrifuged to remove the agarose before use).

4.9.4.3 The membrane was then washed with PBS three times for 5 minutes each. The membranes were dried before being scanned on the infrared imaging system (1-9).

Note: Following synthetic DNA polynucleotide can be used instead of natural DNA

☆ *Poly (dG)-poly(dC) and poly (dA)-poly(dT), in which each strand of the dsDNA is a homopolymer; and*

☆ *Poly(dG-dC)-poly (dG-dC) and poly(dA-dT)–poly (dA-dT), in which each strand of the dsDNA is a heteropolymer of alternating bases.*

References

1. Summersn FA, Mason R P and Ehrenshaft M (2013). Development of immune-blotting techniques for DNA radical detection. *Free Rad. Biol. Med.* 5664–5671.

2. Breen A P and Murphy J A (1995). Reactions of oxyl radicals with DNA. *Free Radic. Biol. Med.* 18, 1033–1077.

3. Imlay J A (2008). Cellular defenses against superoxide and hydrogen peroxide. *Annu. Rev. Biochem.* 755–776.

4. Gunther M R, Hanna P M, Mason R P and Cohen M S (1995). Hydroxyl radical formation from cuprous ion and hydrogen peroxide: A spin-trapping study. *Arch. Biochem. Biophys.* 316, 515–522.

5. Ramirez D C, Gomez-Mejiba S E and Mason R P (2006). Immuno-spin trapping of DNA radicals. *Nat. Methods.* 3, 123–127.

6. Cadet J, Douki T and Ravanat J L (2011). Measurement of oxidatively generated base damage in cellular DNA. *Mutat. Res.* 711, 3–12.

7.   Masek T, Vopalensky V, Suchomelova P and Pospisek M (2005). Denaturing RNA electrophoresis in TAE agarose gels. *Anal. Biochem.* 336, 46–50.

8.   Williams D L (1990). The use of a PVDF membrane in the rapid immobilization of genomic DNA for dot blot hybridization analysis. *Biotechniques.* 8, 14–15. [

9.   Hicks D A and Vecoli C (1987). The use of PVDF membrane in the alkaline transfer of DNA. *Biotechniques.* 5, 206.

Protocol No. 4.10: Evaluation of Radiation Induced DNA Oxidation and its Protection by Rdioprotective Agent using ELISA Method

Assay Requirement

DNA, DMPO, PBS (pH 7.2), copper (II) chloride, potassium phosphate, sodium phosphate, potassium chloride, sodium chloride, sodium acetate, DTPA, ice-cold ethanol, Tris-EDTA buffer (pH 8.0), opaque ELISA plates, reaction-Bind DNA coating solution, dimethylsulfoxide, glyoxal, sodium phosphate, casein, monoclonal anti-DMPO antibody, goat anti-mouse IgG–HRP, chicken anti-DMPO nitrone IgY, rabbit anti-chicken IgY(H+L)-HRP antibodies, LumiGLO substarte/ BCIP-NBT substrate.

4.10.1 Assay Procedure

4.10.1.1 Following experimental groups were formed

    (i) Control DNA; (non irradiated DNA treated with DMPO in PBS)

    (ii) DNA irradiated either with fix dose (>3kGy) or different doses in increasing order (*i.e.* 3, 5,7,10 kGy) of gamma radiation in absence of radioprotective compoundin PBS.

    (iii) DNA irradiated either with fix dose (>3kGy) or different doses in increasing order (*i.e.* 3, 5,7,10 kGy) of gamma radiation in presence of radioprotective compound in PBS.

4.10.1.2 Followed by irradiation, copper (II) chloride in PBS and DMPO in PBS (composition: 2mM potassium phosphate, 8mM sodium phosphate, 2.7mM potassium chloride, and 137mM sodium chloride, pH7.4) was added in all sets of reaction mixtures.

4.10.1.3 Reaction mixture was incubated for 30 minutes to 1h at 37 °C temperature.

4.10.1.4 Followed by incubation, DTPA was added to a final concentration of 1mM to terminate the reaction.

4.10.1.5 The DNA was then precipitated with 1/10 volume of 3M sodium acetate, (pH 5.2) and 2 volumes of ice-cold ethanol and incubated for 10 minutes at room temperature (*Note: that at 4 °C,DMPO will also be precipitated*).

4.10.1.6 The DNA was centrifued at 13000 rpm for 15 minutes at RT, washed with 70 per cent (v/v) ethanol and re-dissolved in 10mMTris, 1mM EDTA buffer (pH8.0).

4.10.2 DNA-DMPO Adduct Detection by ELISA

4.10.2.1 DNA (0.5µg) added to the white (opaque) ELISA plates and 200µlof reaction-Bind DNA coating solution was added.

4.10.2.2 The DNA was then allowed to bind to the ELISA plate at room temperature overnight in the dark with gentle agitation.

4.10.2.3 The plate was than washed with PBS.

(Please Note: that If DNA under evaluation was glyoxalated, it should be treated with 66 per cent (v/v) dimethylsulfoxide,1M glyoxal and 1.5 mM sodium phosphate at 37°C for1h).

4.10.2.4 The plate was then washed withPBS and blocked with 1 per cent (w/v) of casein in PBS(pH 7.4).

4.10.2.5 The DMPO nitrone adducts formed were detected either with 5 µg/ml mouse monoclonal anti-DMPO antibody and1:100 goat anti-mouse IgG–HRP

or

4.10.2.6 With 10µg/ml chicken anti-DMPO nitrone IgY and 1:20,000 rabbit anti-chicken IgY(H+L)–HRP antibodies (each 1h incubations at 37 °C).

4.10.2.7 The plate was than washed three times with PBS.

4.10.2.8 The LumiGLO substrate/(BCIP-NBT substrate) was added and incubated at room temperature.

4.10.2.9 Luminescence/colourimetric measurement was done using ELISA plate reader (1-10).

Note: Following synthetic polynucleotides can be used instead of natural DNA template:

☆ Poly (dG)-poly(dC) and poly (dA)-poly(dT), in which each strand of the dsDNA is a homopolymer; and

☆ Poly(dG-dC)-poly (dG-dC) and poly(dA-dT)–poly (dA-dT), in which each strand of the dsDNA is a heteropolymer of alternating bases.

References

1. Summersn F A, Mason R P and Ehrenshaft M (2013). Development of immunoblotting techniques for DNA radical detection. *Free Rad. Biol. Med.* 5664–5671.

2. Kojima C, Ramirez D C, Tokar E J, Himeno S, Drobna Z, Styblo M, Mason R P, Waalkes M P (2009). Requirement of arsenic biomethylation for oxidative DNA damage. *J. Natl. Cancer Inst.* 101,1670–1681.

3. Detweiler C D, Deterding L J, Tomer K B, Chignell C F, Germolec D and Mason R P (2002). Immunological identification of the heart myoglobin radical formed by hydrogen peroxide. *Free Radic. Biol. Med.* 33, 364–369.

4. Ramirez D C, Gomez-Mejiba S E and Mason R P (2006). Immuno-spin trapping of DNA radicals. *Nat. Methods.* 3, 123–127.

5. Ramirez D C, Gomez-Mejiba S E and Mason R P (2007). Immuno-spin trapping analyses of DNA radicals. *Nat Protoc*. 2, 512–522.

6. Bhattacharjee S, Deterding L J, Chatterjee S, Jiang J, Ehrenshaft M, Lardinois O, Ramirez D C, Tomer K B and Mason R P (2011). Site-specific radical formation in DNA induced by Cu(II)-H_2O_2 oxidizing system, using ESR, immuno-spin trapping, LC-MS, and MS/MS. *Free Radic. Biol. Med*. 50, 1536–1545.

7. Yin B, Whyatt R M, Perera F P, Randall M C, Cooper T B and Santella R M (1995). Determination of 8- hydroxydeoxyguanosine by an immuno affinity chromatography-monoclonal antibody-based ELISA. *Free Radic. Biol. Med*. 18, 1023–1032.

8. Kuhn K M, DeRisi J L, Brown P O and Sarnow P (2001). Global and specific translational regulation in the genomic response of Saccharomyces cerevisiae to a rapid transfer from a fermentable to a nonfermentable carbon source. *Mol. Cell. Biol*. 21, 916–927.

9. Hanna P M, Chamulitrat W and Mason R P (1992). When are metal ion-dependent hydroxyl and alkoxyl radical adducts of 5, 5-dimethyl-1-pyrroline N-oxide artifacts? *Arch. Biochem. Biophys*. 296, 640–644.

10. Makino K, Hagiwara T, Hagi A, Nishi M and Murakami A (1990). Cautionary note for DMPO spin trapping in the presence of iron ion. *Biochem. Biophys. Res. Commun*. 172, 1073–1080.

Protocol No. 4.11: Evaluation of DNA Repair Enhancing Capacity in Terms of 8-OxoG DNA Adduct Elimination from the Damage DNA Strand by Radioprotective Drug Pretreatment to Irradiated Cells

Assay Requirement

Tris–HCl (pH6.5), DTT, NaCl, EDTA, Cell culture facility, culture dishes, culture flask, CO_2 incubator, laminar air flow system, ^{32}P labelled 8-oxoG containing oligonucleotide (sequence mentioned in the protocol), proteinase K, SDS, acrylamide, Bis-acrylamide, TEMED, ammonium per sulfate, phosphorimager, gamma irradiator.

4.11.1 Assay Procedure

4.11.1.1 Following treatment groups was prepared for experimental purpose

    i. Control cells (untreated cells)

    ii. Cells treated with different concentrations of radioprotective drug

    iii. Cells irradiated with gamma radiation (2-4 Gy) in absence of radioprotective drug

    iv. Cells irradiated with gamma radiation (2-4 Gy) in presence of radioprotective drug

    After irradiation and drug treatments cells were harvested and a cell free homogenate was prepared in PBS (pH7.2).

4.11.1.2 8-oxoG containing oligonucleotide i.e 5"-CCAGTGAATTCCCGG GG*ATCCGTCGAC-CTGCAGCCAAGCT-3"(G*= 8oxoG) and its complementary strand was synthesised and labelled with ^{32}P at 5' end. This radiolebelled DNA duplex was used as OGG1 substrate in the reaction.

4.11.1.3 300 fmol of the 5'-end labelled double stranded oligonucleotide substrate was incubated with 100 µl of cell free homogenate prepared from the irradiated and radioprotective drug treated cells ($3×10^6$cells) at 37°C for 90 minutes in a volume of 200µl containing reaction buffer [Composition: 50 mM Tris–HCl (pH6.5), 5 mM DTT, 75 mM NaCl and 0.5 mM EDTA].

4.11.1.4 The reaction mixture was incubated with proteinase K (0.45 mg/ml)

4.11.1.5 DNA fragments were analysed using 20 per cent denaturating PAGE. Image analysis of the separated DNA fragment was carried out using phosphor imager or equivalent imaging device.

4.11.1.6 OGG1 activity was estimated by quantication of the radioactivity in the two separatedbands visualized on the 20 per cent denaturating PAGE in relation to that of the 40-mer substrate (300 fmol).

4.11.1.7 The amount of cleaved molecules (fmol) was corresponded to the OGG1 activity of the cells (3×10^6) per 90 minutes (1-10).

References

1. Janssen K, Schlink K, Gotte W, Hippler B, Kaina B and Oesch F (2001). DNA repair activity of 8-oxoguanine DNA glycosylase I (OGG1) in human lymphocytes is not dependent on genetic polymorphism Ser326/Cys326. *Mutation Res.* 486, 207–216.

2. Paz-Elizur T, Sevilya Z, Leitner-Dagan Y, Elinger D, Roisman L and Livneh Z (2008). DNA repair of oxidative DNA damage in human carcinogenesis: Potential application for cancer risk assessment and prevention. *Cancer Lett.* 266(1), 60–72.

3. Maki H and Sekiguchi M (1992). MutT protein specifically hydrolyses a potent mutagenic substrate for DNA synthesis. *Nature.* 355, 273–275.

4. Seeberg E, Eide L and Bjoras M (1995). The base excision repair pathway. *Trends Biochem. Sci.* 20, 391–397.

5. Krokan H E, Nilsen H, Skorpen F, Otterlei M and Slupphaug G (2000). Base excision repair of DNA in mammalian cells. *FEBS Lett.* 476, 73–77.

6. Fortini P and Dogliotti E (2007). Base damage and single-strand break repair: mechanisms and functional significance of short- and long-patch repair sub pathways. *DNA Repair.* 6, 398–409.

7. Boiteux S and Radicella J P (2000). The human OGG1 gene: structure, functions, and its implication in theprocess of carcinogenesis. *Arch. Biochem. Biophys.* 377, 1–8.

8. Bruner S D, Norman D P and Verdine G L (2000). Structural basis for recognition and repair of the endogenous mutagen 8-oxoguanine in DNA. *Nature.* 403-403

9. Minowa O, Arai T, Hirano M, Monden Y, Nakai S, Fukuda M, Itoh M, Takano H, Hippou Y,Aburatani H, Masumura K, Nohmi T, Nishimura S and Noda T (2000). Mmh/Ogg1 gene inactivation results in accumulation of 8-hydroxyadenine in mice. *Proc. Natl. Acad. Sci.* 97, 4156–4161.

10. Sakumi K, Tominaga Y, Furuichi M, Xu P, Tsuzuki T, Sekiguchi M, Nakabeppu Y (2003). Oggl knockout associated lung tumorigenesis and its suppression by Mth1 gene disruption. *Cancer Res.* 63, 902–905.

Protocol No. 4.12: Evaluation of 8-OxoG DNA Adduct Repairing Enzyme *i.e.* OGG1 Activity in Irradiated Cells and its Modulation by Radioprotective Drug Pretreatment using RT-PCR Assay

Assay Requirement

Cell culture facilities, culture media specific to the cells to be grown, RNA extraction kits, cDNA synthesis kit, sets of primers (sequence mentioned below), agarose, gel elcetrophoresis system, gamma irradiator.

4.12.1 Assay Procedure

4.12.1.1 Following treatment groups was prepared for experimental purpose

    i. Control cells (untreated cells)

    ii. Cells treated with different concentrations of radioprotective drug

    iii. Cells irradiated with gamma radiation (2-4 Gy) in absence of radioprotective drug

    iv. Cells irradiated with gamma radiation (2-4 Gy) in presence of radioprotective drug

After irradiation and drug treatments, cells were harvested by trypsinization.

4.12.1.2 Total RNA was extracted from the treated cells (5×10^6) using standered procedure/commercially available RNA extraction kits.

4.12.1.3 Residual genomic DNA from the RNA preparation was eliminated by DNAase I treatment.

4.12.1.4 Now purified total RNA was subjected to first-strand cDNA synthesis using commercially available cDNA synthesis kit.

4.12.1.5 While, remaining free RNA from the cDNA prepation was removed by RNAase (4μg) treatment.

4.12.1.6 After that, exons 1–4 of OGG1 gene were amplied using following set of primers:

5-TCCATCCCGTGCCCTCGCTCTG-3

5-GATGGCTCGGGCACTGGCACTCA-3

4.12.1.7 Wheras, exons 4–7 of OGG1 gene were amplied using following set of primers:

5-GCTGGGCCTGGGCTATCGTG-3

5-CTGCGCTTTGCTGGTGGCTCCTG-3

4.12.1.8 PCR procedure was carried out using cDNA (4µl) in a volume of 50 µl containing 2.5 unit Hot Star Taq DNA Polymerase and 50µM dNTPs.

4.12.1.9 Reaction was carried out for 35 cycles using following conditions (at 94°C for 1 minute for denaturation; at 64°C for 1.5 minutes for annealing and at 72°C for 2 minutes for extension).

4.12.1.10 Followed by completion of amplification reaction, PCR products were separated by 1.0 per cent agarose gel electrophoresis (1-10).

References

1. Janssen K, Schlink K, Gotte W, Hippler B, Kaina B and Oesch F (2001). DNA repair activity of 8-oxoguanine DNA glycosylase I (OGG1) in human lymphocytes is not dependent on genetic polymorphism Ser326/Cys326. *Mutation Res.* 486, 207–216.

2. Paz-Elizur T, Sevilya Z, Leitner-Dagan Y, Elinger D, Roisman L and Livneh Z (2008). DNA repair of oxidative DNA damage in human carcinogenesis: Potential application for cancer risk assessment and prevention. *Cancer Lett.* 266(1), 60–72.

3. Maki H and Sekiguchi M (1992). MutT protein specifically hydrolyses a potent mutagenic substrate for DNA synthesis. *Nature.* 355, 273–275.

4. Seeberg E, Eide L and Bjoras M (1995). The base excision repair pathway. *Trends Biochem. Sci.* 20, 391–397.

5. Krokan HE, Nilsen H, Skorpen F, Otterlei M and Slupphaug G (2000). Base excision repair of DNA in mammalian cells. *FEBS Lett.* 476, 73–77.

6. Fortini P and Dogliotti E (2007). Base damage and single-strand break repair: mechanisms and functional significance of short and long-patch repair sub-pathways. *DNA Repair.* 6, 398–409.

7. Boiteux S and Radicella J P (2000). The human OGG1 gene: structure, functions, and its implication in theprocess of carcinogenesis. *Arch. Biochem. Biophys.* 377, 1–8.

8. Bruner SD, Norman D P and Verdine GL (2000). Structural basis for recognition and repair of the endogenous mutagen 8-oxoguanine in DNA. *Nature.* 403-403

9. Minowa O, Arai T, Hirano M, Monden Y, Nakai S, Fukuda M, Itoh M, Takano H, Hippou Y, Aburatani H, Masumura K, Nohmi T, Nishimura S and Noda T (2000). Mmh/Ogg1 gene inactivation results in accumulation of 8-hydroxyadenine in mice. *Proc. Natl. Acad. Sci.* 97, 4156–4161.

10. Sakumi K, Tominaga Y, Furuichi M, Xu P, Tsuzuki T, Sekiguchi M, Nakabeppu Y (2003). Oggl knockout associated lung tumorigenesis and its suppression by Mth1 gene disruption. *Cancer Res.* 63, 902–905.

Section 5

Radiation Induced Mitochondrial DNA Damage and its Modulation by Radioprotective Drug Treatment

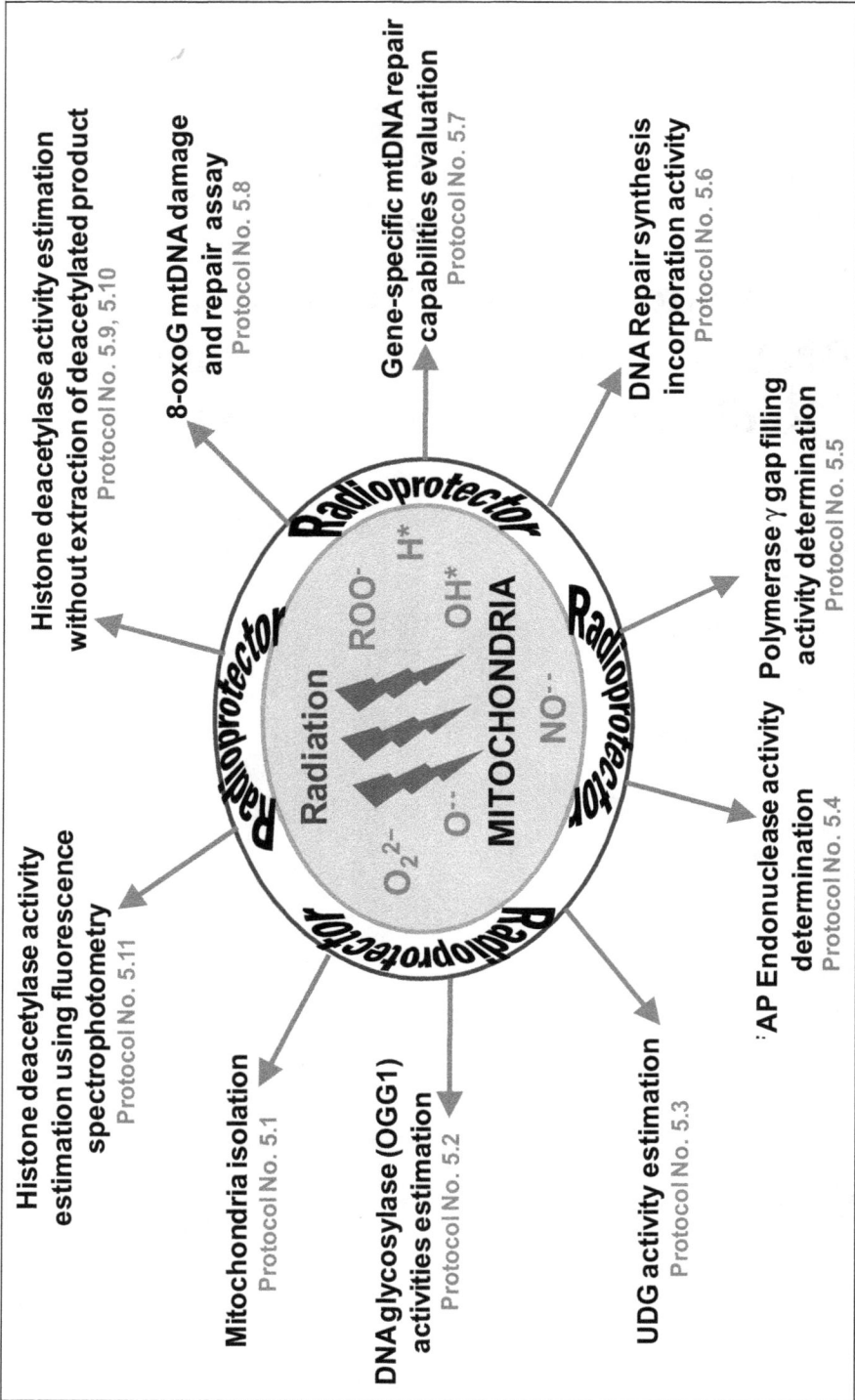

Figure 13: Schematic Summary of Protocols Associated with Determination of Gamma Radiation Induced Damage to Mitochondrial DNA.

Protocol No. 5.1: Isolation of Mitochondria from different Tissues and Cells

Introduction

Reactive oxygen species (ROS) generated during normal cell metabolism is the prime cause of mitochondrial DNA (mtDNA) damage. The common mtDNA lesions induced by ROS are belongs to base modifications, such as the ubiquitous 8-oxoguanine (8-oxoG) lesion.Although, DNA base damage and strand breaks may also induced by ROS. Various diseases and are associated with mtDNA damage and mutations. Most of the DNA lesions are generally repaired by the base excision repair (BER) pathway. BER involve DNA glycosylases, AP endonuclease, DNA polymerase (POLγ in mitochondria) and DNA ligase in a similar fashion as nuclear BER. In view of that, DNA damage and repair methodology have significant impact on the screening of protective chemical agents against ROS induced DNA damage in irradiated cells or tissue gamma radiation (1-8).

Assay Requirement

MSHE buffer, mannitol, sucrose, HEPES, EGTA, EDTA, spermidine, spermine, leupeptin, benzamidine, DTT, glass–teflon Potterhomogenizer, centrifuge, percoll, protease, KCl buffer, Tris base, Mg_2SO_4, EDTA, ATP buffer, ATP, BSA, digitonin, tissue culture plates, dounce glass-homogenizer, subtilisin, lysis buffer, Tris–HCl, KCl, EDTA, DTT, glycerol, NP-40, PMSF, protease inhibitor cocktail, HEPES, EDTA, glycerol, $MgCl_2$, de-freezer, gamma irradiator.

5.1.1 Assay Procedure

5.1.1.1 Isolation of Mitochondria from Liver, Kidney and Testicular Homogenates

5.1.1.1.2 Tissue of interest was transferred to a beaker containing ice cold MSHE buffer (composition: 210 mM mannitol, 70 mM sucrose, 10 mM HEPES, 1 mM EGTA, 2 mM EDTA, 0.75 mM spermidine, 0.15 mM spermine, 2µg/ml leupeptin, 5 mM DTT, 2 µM benzamidine, pH 7.4).

5.1.1.1.3 Tissue was minced with scissors and washed in MSHE buffer. Minced tissue was transferred to a glass–teflon Potter–homogenizer and homogenized until a smooth homogenate produced.

5.1.1.1.4 Tissue homogenate was subjected to low speed spin (LSS, 1000xg for 10 minutes) so that unbroken cells and the nuclear fraction can be precipitated.

5.1.1.1.5 The supernatant is transferred to another tube and a high speed spin (HSS, 10,000xg for 10 minutes) was performed.

5.1.1.1.6 The resulted mitochondrial pellet obtained was washed with MSHE followed by another HSS (10,000xg for 10 minutes).

5.1.1.1.7 The pellet was then re-suspended in 500 µl MSHE buffer and layered on the top of a gradient mixture of 50 per cent percoll-50 per cent 2 X MSHE and spun at 50,000xg.

5.1.1.1.8 The mitochondrial fraction was found to appears as a band at the one-third of the distance from the top of the centrifuge tube.

5.1.1.1.9 The mitochondria was removed and washed free of percoll by the addition of 10 volumes of MSHE buffer.

5.1.1.1.10 The pure mitochondria were precipitated in a 3000xg spin and mitochondrial enzymes can be purified.

5.1.1.2 Isolation of the Mitochondrial Fraction from Brain Tissue

5.1.1.2.1 The brain tissue was added to ice cold MSHE buffer and minced with scissors.

5.1.1.2.2 After MSHE buffer wash, the tissue was homogenized and an LSS (1000xg for 10 min) was performed to precipitate the nuclear fraction.

5.1.1.2.3 The supernatant contains mitochondria was further subjected to High Speed Spin (10,000xg for 10 minutes) to precipitate the mitochondria.

5.1.1.2.4 A portion of these mitochondria are contained in synaptosomal vesicles. that can be disrupt by addition of digitonin, however, this detergent has been shown to disrupt the outer mitochondrial membrane.

5.1.1.2.5 After a second HSS (10,000xg for 10 minutes) the mitochondria can be further purified on a percoll gradient as described above.

5.1.1.3 Isolation of the Mitochondrial Fraction from Skeletal Muscle

5.1.1.3.1 Mitochondria in skeletal muscle are often isolated using slightly different method than mitochondria from other tissues due to the morphological properties of muscle tissue.

5.1.1.3.2 There is two main differences separate the modified procedure. First, the nuclear fraction is much smaller in skeletal muscle than many other tissues, thus nuclear contamination becomes less of a problem.

5.1.1.3.3 Second, skeletal muscle tissue is organized in bundles of fibers bound together by different layers of connective tissue (endomysium, perimysium and epimysium) and mitochondria are organized very stringently in a crystallized pattern bound tightly to the cytoskeleton.

5.1.1.3.4 Separation of subcellular compartments can be facilitated by the addition of a protease prior to, or during the homogenization.

5.1.1.3.5 It was observed that addition of 0.1 mg/ml nagarse (or a comparable protease) during homogenization to be especially useful to isolate mitochondria from heart or skeletal muscle.

5.1.1.3.6 In brief, skeletal muscle tissue was kept in ice cold KCl buffer (100 mM KCl, 50 mM Tris base, 5 mM Mg_2SO_4, 1 mM EDTA) and mincing with scissors. The liquid is removed and 5 ml of ATP buffer (KCl with 1 mM ATP, 0.5 per cent BSA) was added.

5.1.1.3.7 Tissue and buffer was transferred to a glass Teflon Potter–Elvehjem homogenizer and homogenized until a smooth homogenate was obtained.

5.1.1.3.8 If nagarse was added the homogenate it should be spun at 9500xg for 9 minutes.

5.1.1.3.9 The supernatant removed and the pellet was resuspended in ATP buffer.

5.1.1.3.10 An LSS (1000xg for 10 minutes) was performed and the supernatant transferred to a new tube.

5.1.1.3.11 An HSS (5400xg for 10 minutes) is performed and the pellet was re-suspended and further subjected to another HSS (6700xg for 10 minutes).

5.1.1.3.12 The resulting mitochondrial pellet can now be used for enzyme isolation.

5.1.1.4 Preparation of mitochondrial extracts

5.1.1.4.1 Followed by mitochondrial fraction isolation, a crude extract of the mitochondrial fraction needs to be prepared for *in vitro* assays.

5.1.1.4.2 In brief, 100 µL of lysis buffer (composition:10 mM Tris–HCl pH 7.8, 400 mM KCl, 1 mM EDTA, 1 mM DTT, 20 per cent glycerol, 0.1 per cent NP-40, 0.25 mM PMSF, 1x protease inhibitor cocktail) was added to the mitochondrial pellet and incubated at 1.5h d at 4 °C.

5.1.1.4.3 The lysate was then sonicated at 5W for 5second 5 times in ice, with 30 second intervals and spun at 130,000xg to precipitate contaminating DNA and membranes.

5.1.1.4.4 The supernatant containing the purified proteins was dialyzed overnight against the dialysis buffer (25 mM HEPES, 100 mM KCl, 1 mM EDTA, 17 per cent glycerol,12 mM $MgCl_2$).

5.1.1.4.5 Followed by dialysis, the lysate was spun at 16,000xg for 10 minutes to precipitate any non-soluble fraction.

5.1.1.4.6 Consequently, the supernatant was aliquoted in different tubes. Protein concentration was measured before assay performed.Aliquoted mitochondrial suspension can be stored at -80 °C for several months without significant loss of enzyme activity; however, successive freeze–thaw cycles in the same aliquot should be avoided (1-8).

References

1. Maynard S, de Souza-Pinto N, Scheibye-Knudsen M and Bohr V A (2010). Mitochondrial base excision repair assays. *Methods*. 51, 416–425.

2. Desjardins P, de Muys JM and Morais R (1986). An established avian fibroblast cell line without mitochondrial DNA. *Somat. Cell Mol. Genet.* 12(2), 133-9.

3. Hamilton M L, Guo Z, Fuller C D, Van Remmen H, Ward W F, Austad S N, Troyer D A, Thompson I and Richardson A (2001). A reliable assessment of 8-oxo-2-deoxyguanosine levels in nuclear and mitochondrial DNA using the sodium iodide method to isolate DNA. *Nucleic Acids Res.* 29(10), 2117-26.

4. de Souza-Pinto N C, Wilson D M, Stevnsner T V and Bohr V A (2008). Mitochondrial DNA, base excision repair and neurodegeneration. *DNA Repair (Amst).* 7(7), 1098-1099.

5. Wilson G L, Beecham E J, Stevnsner T, Wassermann K and Bohr V A (1992). Repair of mitochondrial DNA after various types of DNA damage in Chinese hamster ovary cells. *Carcinogenesis.* 13(11), 1967-1973.

6. Pinz K G and Bogenhagen D F (1998). Efficient repair of abasic sites in DNA by mitochondrial enzymes. *Mol Cell Biol.* 18(3), 1257-1265.

7. Thyagarajan B, Padua R A and Campbell C (1996). Mammalian mitochondria possess homologous DNA recombination activity. *J Biol. Chem.* 271(44), 27536-27543.

8. Pallotti F and Lenaz G (2007). Isolation and sub-fractionation of mitochondria from animal cells and tissue culture lines. *Methods Cell Biol.* 80. 3-44.

Protocol No. 5.2: Determiation of Effect of Radioprotective Drug on DNA Glycosylase (OGG1 Activity) Activities in Mitochondria of Irradiated Cells

Background

DNA glycosylase catalyze initial reaction of the Base Excision Repair (BER) pathway. DNA glycosylase accelerate releasing of the damaged base through cleavage of the N-glycosyl bond that joint the nitrogen base to the ribose moiety of the nucleotide chain. The DNA glycosylase assay measured enzyme activity associated with the cleavage of a single-lesion-containing oligonucleotide.The enzyme activity can be calculated as the amount of cleaved substrate relative to the total amount of substrate in the lane.OGG1 incision activity can be measured using oligonucleotide substrates containing single lesions recognized by this enzyme. 8-oxodG is one of the major in vivosubstrates for this enzyme (1-13).

Assay Requirement

HEPES–KOH buffer (pH 7.4), EDTA, KCl, glycerol, Triton X-100, DTT, protease inhibitors, KCl, glycerol, ^{32}P-labeled duplex oligonucleotide (sequence mentioned below), SDS, proteinase K, glycogen, ethanol, ammonium acetate, formamide dye, EDTA, bromophenol blue, xylene cyanol.

5.2.1 Assay Procedure

5.2.1.1 Cells or animals were divided in following four groups:

Gp 1: Control; untreated cells/animals

Gp 2: Irradiated cells/animals

Gp 3: Radioproductive drug treated cells/animals

Gp 4: Irradiated cells/animals pretreated with radioproductive drug.

5.2.1.2 Followed by all treatments, mitochondria from treated cells or animal tissues were isolated using specific protocols (refer Protocol No. 5.1).

5.2.1.3 Protein concentration of the mitochondrial lysates was measured and set as per required in each assay with the following buffer: 20 mM HEPES–KOH (pH 7.4), 1 mM EDTA, 100 mM KCl, 25 per cent glycerol (v/v), 0.015 per cent Triton X100, 5 mM DTT and protease inhibitors.

5.2.1.4 Incision reactions (20 µl) containing following ingredients was performed: 40 mM HEPES–KOH (pH 7.4)

5 mM EDTA

1 mM DTT

75 mM KCl

10 per cent glycerol and

95 fmol of ^{32}P-labeled duplex oligonucleotide:

50-GAA CGA CTG T(**OG**)A CTT GAC TGC TAG TGA T

30-CTT GCT GAC ACT GAA CTG ACG ATG ACT A

The reactions were initiated by adding the mitochondrial lysates isolated from cells/animal tissue of different treatment groups.

[Note: Mitochondria from animal tissues, such as liver and brain have higher OGG1 activity than mitochondria from cultured cells, thus require less lysate in the assay. Typically, OGG1 activity can be measured in the range of 10–100µg protein per reaction; however, a concentration curve with increasing amounts of lysate must be prepared].

5.2.1.5 It is important to carefully calibrate the amount of protein for the assay. Too much protein may resulted in non-specific degradation of the substrates.

5.2.1.6 The reactions are then incubated at 32 °C for 12–16 h.The reaction was terminated by addition of 1 µl each of the following: proteinase K (5 mg/ml) and 10 per cent SDS and reaction mixture incubated at 55 °C for 30 minutes again.

5.2.1.7 The DNAs was then ethanol-precipitated by the addition of 1µg of glycogen, 4µl of 11M ammonium acetate and 63µl ethanol, pelleted, dried and suspended in formamidedye (90 per cent deionized formamide, 1 mM EDTA, 0.1 per cent bromophenolblue, 0.1 per cent xylene cyanol).

5.2.2 Electrophoretic Analysis of the Base Excision Repair Product

5.2.2.1 The assay described above requires resolving the oligonucleotide substrates under denaturing conditions.

5.2.2.2 20 per cent polyacrylamide gels containing 7M urea, in 1x Tris–borate/EDTA (TBE) buffer was used.

5.2.2.3 The acrylamide solution was prepared by diluting a commercially available acrylamide/bis-acrylamide solution (37:1) in H$_2$O and adding 7M urea, and 0.1 volume of 10x TBE.

5.2.2.4 After completely dissolving the urea, the solution was filtered.

5.2.2.5 This stock can be stored up to 2 months in the refrigerator, protected from light.

5.2.2.6 For a better resolution, used 1mm gels of at least 16cm length.

5.2.2.7 The samples was loaded and the gels resolved in TBE buffer at 16W, for 1h.

5.2.2.8 The gels can be dried, but if the radioactive signal is strong enough, they can be exposed without drying.

5.2.2.9 The signals are then acquired using a Phospho-Imager (Molecular Dynamics) or equivalent scanner.

5.2.2.10 The radioactivity in the lanes was quantified using the Image Quant software Incision activities was quantified as the amount of radioactivity in the product lane relative to the total radioactivity in the lane.

5.2.2.11 Gap filling and incorporation activity was calculated as the amount of radioactivity in the product lane, normalized by the control labeled substrate, relative to the incorporation of the control samples, which are set as 100 per cent activity (1-13).

References

1. Maynard S, de Souza-Pinto N, Scheibye-Knudsen M and Bohr V A (2010). Mitochondrial base excision repair assays. *Methods*. 51, 416–425.

2. Persinger R L, Melamede R, Bespalov I, Wallace S, Taatjes D J and Janssen-Heininger Y (2001). Imaging techniques used for the detection of 8-oxoguanine adducts and DNA repair proteins in cells and tissues. *Exp. Gerontol*. 36(9),1483-1494.

3. Beckman K B and Ames B N (1999). Endogenous oxidative damage of mtDNA. *Mutat Res*. 424(1-2), 51-58.

4. Beckman K B and Ames B N (1996).Detection and quantification of oxidative adducts of mitochondrial DNA. *Methods Enzymol*. 264,442-453.

5. ESCODD -European Standards Committee on Oxidative DNA Damage (2002). Comparative analysis of baseline 8-oxo-7,8-dihydroguanine in mammalian cell DNA by different methods in different laboratories: an approach to consensus. *Carcinogenesis*. 23(12), 2129-2133.

6. Ohno M, Oka S and Nakabeppu Y (2009). Quantitative analysis of oxidized guanine, 8-oxoguanine in mitochondrial DNA by immunofluorescence method. *Methods Mol Biol*. 554,199-212.

7. de Souza-Pinto N C, Maynard S, Hashiguchi K, Hu J, Muftuoglu M and Bohr V A (2009). The recombination protein RAD52 cooperates with the excision repair protein OGG1 for the repair of oxidative lesions in mammalian cells. *Mol Cell Biol*. 29(16), 4441-4454.

8. Struthers L, Patel R, Clark J and Thomas S (1998).Direct detection of 8-oxodeoxyguanosine and 8-oxoguanine by avidin and its analogues. *Anal Biochem*. 255(1), 20-31.

9. Soultanakis R P, Melamede R J, Bespalov I A, Wallace S S, Beckman K B, Ames B N, Taatjes D J and Janssen-Heininger Y M (2000). Fluorescence detection of 8-oxoguanine in nuclear and mitochondrial DNA of cultured cells using a recombinant Fab and confocal scanning laser microscopy. *Free Radic. Biol. Med*. 28(6), 987-998.

10. Ide H, Kow Y W, Chen B X, Erlanger B F and Wallace S S (1997). Antibodies to oxidative DNA damage: characterization of antibodies to 8-oxopurines. *Cell Biol Toxicol.* 13(6), 405-417.

12. Singh N P, McCoy M T, Tice R R and Schneider E L (1988). A simple technique for quantitation of low levels of DNA damage in individual cells. *Exp. Cell Res.* 175(1), 184-191.

13. Collins A R, Duthie S J and Dobson V L (1993). Direct enzymic detection of endogenous oxidative base damage in human lymphocyte DNA. *Carcinogenesis.* 14(9),1733-1735.

14. Vlachogianni T and Fiotakis C (2009). 8-hydroxy-2' -deoxyguanosine (8 OHdG): A critical biomarker of oxidative stress and carcinogenesis. *J Environ Sci Health C Environ Carcinog Ecotoxicol Rev.* 27(2),120-139.

Protocol No. 5.3: Determination of UDG Activity in Mitochondria of Irradiated Cells and its Modulation by Radioprotective Drug Pretreatment

Assay Requirement

Cell culture plates, HEPES–KOH buffer (pH 7.4), EDTA, KCl, glycerol, Triton X-100, DTT and protease inhibitors, NaCl, ^{32}P-labeled duplex oligonucleotide (sequence mentioned below), NaOH, Tris–borate/EDTA (TBE) buffer, acrylamide, bis-acrylamide, urea, deionized water, Phospho-Imager/equivalent scanner, gamma irradiator, cell culture facility.

5.3.1 Assay Procedure

5.3.1.1 Cells or animals were divided in following four groups:

Gp 1: Control; untreated cells/animals

Gp 2: Irradiated cells/animals

Gp 3: Radioprotective drug treated cells/animals

Gp 4: Irradiated cells/animals pretreated with radioprotective drug

5.3.1.2 Followed by all treatments, mitochondria from treated cells or animal tissues were isolated using specific protocols mentioned above (No. 5.1).

5.3.1.3 Protein concentration of the mitochondrial lysates was measured and set as per required in each assay with the following buffer: 20 mM HEPES–KOH (pH 7.4), 1 mM EDTA, 100 mM KCl, 25 per cent glycerol (v/v), 0.015 per cent Triton X-100, 5 mM DTT and protease inhibitors.

5.3.1.4 Uracil incision activity can be measured using a 30-mer double stranded oligonucleotide containing a single uracil (U) at position 12. Incision reactions (20 µl) contain following components:

70 mM HEPES–KOH (pH 7.4)

5 mM EDTA

1 mM DTT

75 mM NaCl

10 per cent glycerol and

50 fmol of ^{32}P-labeled duplex oligonucleotide *i.e.*:

50-ATA TAC CGC GG(**U**) CGG CCG ATC AAG CTT ATT

30-TAT ATG GOG CC G GCC GGC TAG TTC GAA TAA

5.3.1.5 The reaction was initiated by adding the mitochondrial lysate and incubated for 1h at 37 °C. In general, UDG activity is much higher than OGG1, so, lower amounts of protein lysate (1-10µg of protein)are needed.

5.3.1.6 After completion, the reactions terminated and DNA was processed similarly as mentioned in the OGG1 assay above (refer Protocol No. 5.2).

5.3.1.6 Because UDG does not have an associated AP-lyase activity, the product of the reaction will be an abasic-site containing oligonucleotide which has to be converted into a single-strand break.

5.3.1.7 For that 50 mM NaOH should be added to the samples and resuspended in the loading dye and incubated for 15 minutes at 75°C before loading into the gel.

5.3.2 Electrophoresis

5.3.2.1 The assays described above needed resolving the oligonucleotide substrates under alkaline (denaturing) conditions.

5.3.2.2 20 per cent polyacrylamide gels containing 7M urea, in 1x Tris–borate/EDTA (TBE) buffer was used.

5.3.2.3 The acrylamide solution was prepared by diluting a commercially available acrylamide/bis-acrylamide solution (37:1) in H_2O and adding 7M urea, and 0.1 volume of 10x TBE.

5.3.2.4 Followed by completely dissolving the urea, the solution was filtered and can be stored in 4°C for 2 months provided protected from the light.

5.3.2.5 For a better resolution, used 1 mm gels of at least 16 cm length.

5.3.2.6 The samples were loaded and the gels resolved in TBE buffer at 16W for 1h.

5.3.2.7 The gels can be dried, but if the radioactive signal is strong enough, they can be exposed without drying.

5.3.2.8 The signals are then acquired using a Phospho-Imager or equivalent scanner.

5.3.2.9 The radioactivity in the lanes is quantified using the Image Quant software.

5.3.2.10 Incision activities are quantified as the amount of radioactivity in the product lane relative to the total radioactivity in the control lane (1-11).

Note: Other DNA Glycosylases

The activities of other DNA glycosylases, like NTH and NEIL1, can be assayed similarly using similar reactions and assay procedures. The substrates need to be chosen according to the substrate specificity of each enzyme. i.e. thymine–glycol-containing oligonucleotide can be used to assay for NTH and a Fapy-Guanine containing substrate for NEIL1. Reaction and salt conditions can be optimized independently.

References

1. Maynard S, de Souza-Pinto N, Scheibye-Knudsen Mand Bohr V A (2010). Mitochondrial base excision repair assays. *Methods*. 51, 416–425.

2. Souza-Pinto N C, Croteau D L, Hudson E K, Hansford R G and Bohr V A (1999). Age-associated increase in 8-oxo-deoxyguanosine glycosylase/AP lyase activity in rat mitochondria. *Nucleic Acids Res*. 27(8), 1935-1942.

3. Dobson A W, Xu Y, Kelley M R, LeDoux S P and Wilson G L (2000). Enhanced mitochondrial DNA repair and cellular survival after oxidative stress by targeting the human 8- oxoguanine glycosylase repair enzyme to mitochondria. *J Biol Chem*. 275(48), 37518-37523.

4. Hartwig A, Dally H and Schlepegrell R (1996). Sensitive analysis of oxidative DNA damage in mammalian cells: Use of the bacterial Fpg protein in combination with alkaline unwinding. *Toxicol Lett*. 88(1-3), 85-90.

5. Trapp C, McCullough A K and Epe B (2007). The basal levels of 8-oxoG and other oxidative modifications in intact mitochondrial DNA are low even in repair-deficient (Ogg1(-/-)/Csb(-/-) mice. *Mutat. Res*. 625(1-2), 155-63.

6. Pflaum M, Will O, Mahler H C and Epe B (1998). DNA oxidation products determined with repair endonucleases in mammalian cells: types, basal levels and influence of cell proliferation. *Free Radic Res*. 29(6), 585-594.

7. Collins A R, Duthie S J and Dobson V L (1993). Direct enzymic detection of endogenous oxidative base damage in human lymphocyte DNA. *Carcinogenesis*. 14(9), 1733-1735.

8. Yin B, Whyatt R M, Perera F P, Randall M C, Cooper T B and Santella R M (1995). Determination of 8-hydroxydeoxyguanosine by an immunoaffinity chromatography-monoclonal antibody-based ELISA. *Free Radic Biol Med*. 18(6), 1023-1032.

9. Struthers L, Patel R, Clark J and Thomas S (1998). Direct detection of 8-oxodeoxyguanosine and 8-oxoguanine by avidin and its analogues. *Anal Biochem*. 255(1), 20-31.

10. Domena JD and Mosbaugh DW (1985). Purification of nuclear andmitochondrial uracil-DNA glycosylase from rat liver. Identification oftwo distinct subcellular forms. *Biochemistry*. 24, 7320–7328.

11. Lindahl T, Ljungquist S, Siegert W, Nyberg B and Sperens B (1977). DNA N glycosidases: properties of uracil-DNA glycosidase from *Escherichia coli*. *J. Biol. Chem*. 252, 3286–3294.

Protocol No. 5.4: Determination of AP Endonuclease Activity in Mitochondria of Irradiated Cells and its Modulation by Radioprotective Drug

Assay Requirement

Cell cultute plates, HEPES–KOH buffer (pH 7.4), KCl, BSA, $MgCl_2$ glycerol, Triton X-100, ^{32}P-labeled duplex oligonucleotide containg tetrahydrofuran (THF) at 11 position (sequence mentioned below), heating oven, incubator, urea, Tris–borate/EDTA (TBE) buffer, acrylamide, bis-acrylamide, deionized water, Phospho-Imager or equivalent scanner, gamma irradiator and cell culture facility.

5.4.1 Assay Procedure

5.4.1.1 In higher animals particularly in mammals, AP endonuclease 1 (APE1) is the major abasic endonuclease which catalyses DNA repair reaction, both in the nucleus and in the mitochondria.The assay is similar to the glycosylase incision activity assay, with only difference that the substrate should now contains either a true abasic site, generated via a class I DNA glycosylase-catalyzed reaction or an abasic site analogue, such as tetrahydrofuran (THF).

5.4.1.2 Abasic site analogue, such as tetrahydrofuran (THF) is suitable for laboratory analysis as it is commercially available and efficiently recognized by APE1.

5.4.1.3 The substrate used should be ^{32}P-labeled and annealed as described earlier, with the small modification that the annealing is performed by heating up the samples upto 75 °C to avoid non-enzymatic cleavage at the abasic site.

5.4.1.4 APE1 incision activity could be measured using a 28-mer oligonucleotide containing the THF analogue at position 11.

5.4.1.5 Cells or animals were divided in following four groups:

Gp 1: Control; untreated cells/animals

Gp 2: Irradiated cells/animals

Gp 3: Radioprotective Drug treated cells/animals

Gp 4: Irradiated cells/animals pretreated with radioprotective drug.

Followed by all treatments, mitochondria from treated cells or animal tissues were isolated using specific protocols mentioned above.

5.4.1.6 The lysate samples are diluted in 10 mM HEPES–KOH (pH 7.4) containing 100 mM KCl. The reactions mixture (10µl) was set with following components:

25 mM HEPES–KOH (pH 7.4)

25 mM KCl

0.1 mg/mL BSA

5 mM $MgCl_2$

10 per cent glycerol

0.05 per cent Triton X-100 and

10 fmol of ^{32}P-labeled duplex oligonucleotide.

50-GAA CGA CTG T (**F**) A CTT GAC TGC TAC TGA T

30-CTT GCT GAC ACT GAA CTG ACG ATG ACT A

5.4.1.7 The protein concentration for the APE1 activity reactions is even lower than that of UDG, little as 25 ng protein per reaction could be enough.

5.4.1.8 The reactions were incubated at 37 °C for 30 minutes, and terminated by the addition of formamide dye and heating at 90 °C for 10 minutes.

5.4.1.9 Samples are resolved, visualized and analyzed with the help of electrophoresis as described below.

5.4.2 Electrophoresis

5.4.2.1 The assay described above requires denaturing conditions to resolve the oligonucleotide substrates.

5.4.2.2 20 per cent polyacrylamide gels containing 7M urea, in 1x Tris–borate/EDTA (TBE) buffer was used.

5.4.2.3 The acrylamide solution was prepared by diluting a commercially available acrylamide/bis-acrylamide solution (37:1) in H_2O and adding 7M urea, and 0.1 volume of 10x TBE and filtered.

5.4.2.4 Acrylamide stock solution can be stored up to 2 months in the refrigerator in dark.

5.4.2.5 For a better resolution, 1 mm gels of at least 16 cm length should be used.

5.4.2.6 The samples were loaded and the gels resolved in TBE buffer at 16W, for 1h.

5.4.2.7 The gels can be dried, but if the radioactive signal is strong enough, they can be exposed without drying.

5.4.2.8 The signals were acquired using a Phospho-Imager or equivalent scanner.

5.4.2.9 The radioactivity in the lanes was quantified using the Image Quant software.

5.4.2.10 Incision activities were quantified as the amount of radioactivity in the product lane relative to the total radioactivity in the control lane (1-10).

References

1. Maynard S, de Souza-Pinto N, Scheibye-Knudsen M and Bohr V A (2010). Mitochondrial base excision repair assays. *Methods.* 51, 416–425.

2. Chattopadhyay R, Wiederhold L, Szczesny B, Boldogh I, Hazra TK, Izumi T and Mitra S (2006). Identification and characterization of mitochondrialabasic (AP)-endonuclease in mammalian cells. *Nucleic Acids Res.* 34(7), 2067–2076.

3. Ribar B, Izumi T and Mitra S (2004). The major role of humanAP-endonuclease homolog Apn2 in repair of abasic sites in *Schizosaccharomyces pombe. Nucleic Acids Res.* 32, 115–126.

4. Bohr VA, Stevnsner T and de Souza-Pinto NC (2002). Mitochondrial DNA repair of oxidative damage in mammalian cells. *Gene.* 286, 127–134.

5. Tomkinson A E, Bonk R T, Kim J, Bartfeld N and Linn S (1990). Mammalian mitochondrial endonuclease activities specific forultraviolet-irradiated DNA. *Nucleic Acids Res.* 18, 929-935.

6. Tomkinson A E, Bonk R T and Linn S (1988). Mitochondrial endonuclease activities specific for apurinic/apyrimidinic sites in DNA from mouse cells. *J. Biol. Chem.* 263, 12532-12537.

7. Szczesny B and Mitra S (2005). Effect of aging on intracellular distribution of abasic (AP) endonuclease 1 in the mouse liver. *Mech. Ageing Dev.* 126, 1071-1078.

8. Bhakat K K, Izumi T, Yang S H, HazraT K and Mitra S (2003). Role of acetylated human AP-endonuclease (APE1/Ref-1) in regulation of the parathyroid hormone gene. *EMBO J.*, 22, 6299-6309.

9. Izumi T and Mitra S (1998). Deletion analysis of human AP-endonuclease: minimum sequence required for the endonuclease activity. *Carcinogenesis.* 19, 525–527.

10. Hill J W, Hazra T K, Izumi T and Mitra S (2001). Stimulation of human 8-oxoguanine-DNA glycosylase by AP-endonuclease: Potential coordination of the initial steps in base excision repair. *Nucleic Acids Res.* 29, 430–438.

Protocol No. 5.5: Determination of Polymerase γ Gap-filling Activity in Mitochondria of Irradiated Cells and its Modulation by Radioprotective Drug

Assay Requirement

Cell culture plates, a ^{32}P-labeled oligonucleotide (sequences mentioned below), Tris–HCl buffer (pH 7.4), KCl, DTT, MgCl$_2$ glycerol, dCTP, duplex gap oligonucleotide having gap at 16 position, ^{32}P-dCTP, heating oven, incubator, urea, Tris–borate/EDTA (TBE) buffer, acrylamide, bis-acrylamide, deionized water, phospho-Imager or equivalent scanner, gamma irradiator and cell culture facility.

5.5.1 Assay Procedure

5.5.1.1 Mitochondrial DNA POLγ catalyses both end-trimming and nucleotide insertion activities.

5.5.1.2 Single nucleotide gap-filling activity can be measured using a labeled 34-mer duplex oligonucleotide having a single gap at position 16. The assay is measured the incorporation of a radioactive nucleotide during repair process and subsequently quantification of radioactivity incorporated in the full-length oligonucleotide substrate.

5.5.1.3 To optimize the variations in the sample loading in the gels, spiking the samples with a ^{32}P-labeled oligonucleotide of a different size to normalize the radioactivity quantifications.

5.5.1.4 Cells or animals were divided in following four groups:

Gp 1: Control; untreated cells/animals

Gp 2: Irradiated cells/animals

Gp 3: Radioprotective Drug treated cells/animals

Gp 4: Irradiated cells/animals pretreated with radioprotective drug.

Followed by all treatments, mitochondria from treated cells or animal tissues were isolated using specific protocols mentioned above (refer Protocol No. 5.1).

5.5.1.5 The lysate was diluted with 10 mM Tris–HCl (pH 7.4) containing 100 mM KCl.

5.5.1.6 The final reaction mixture (10µl) contained:

50 mM Tris–HCl (pH 7.4)

50 mM KCl

1 mM DTT

5 mM MgCl$_2$

5 per cent glycerol

5 µM dCTP

1 pmol of duplex gap oligonucleotide:

50-CTG CAG CTG ATG CGC OGT ACG GAT CCC CGG GTA C

30-GAC GTC GAC TAC GCG GCA TGC CTA GGG GCC CAT G

4 µCi of a^{32}P-dCTP

5.5.1.7 The amount of protein used in these assays should be calibrated for each sample.Mminimum 1-10 µg protein per reaction was found sufficient to run the assay.

5.5.1.8 Reaction mixture was incubated at 37 °C for 1h and terminated by the addition of formamide dye and heating at 90 °C for 10 minutes.

5.5.1.9 Samples were then resolved using gel electrophoresis as described below.

5.5.2 Electrophoresis

5.5.2.1 20 per cent polyacrylamide gels containing 7M urea, in 1x Tris–borate/EDTA (TBE) buffer was used.

5.5.2.2 The acrylamide solution was prepared by diluting a commercially available acrylamide/bis-acrylamide solution (37:1) in H$_2$O and adding 7M urea and 0.1 volume of 10x TBE.

5.5.2.3 Followed by completely dissolving the urea, the solution was filtered.

5.5.2.4 Stock solution can be stored up to 2 months in the refrigerator, provided protected from the light.

5.5.2.5 1 mm gels of at least 16 cm length should be used for sharp resolution.

5.5.2.6 The samples were loaded and the gels resolved in TBE buffer at 16W for 1h.

5.5.2.7 The signals are then acquired using a Phospho-Imager or equivalent scanner.

5.5.2.8 The radioactivity in the lanes is quantified using the Image Quant software.

5.5.2.9 Gap filling activity can be calculated as the amount of radioactivity in the product lane, subtracted from the control labeled substrate (1-6).

References

1. Maynard S, de Souza-Pinto N, Scheibye-Knudsen M and Bohr V A (2010). Mitochondrial base excision repair assays. *Methods*. 51, 416–425.

2. Liu P, Qian L, Sung J S, de Souza-Pinto N C, Zheng L, Bogenhagen D F, Bohr V A, Wilson D M, Shen B and Demple B (2008). Removal of oxidative DNA damage via FEN1-dependent long-patch base excision repair in human cell mitochondria. *Mol Cell Biol*. 28(16), 4975-4987.

3. Szczesny B, Tann A W, Longley M J, Copeland W C and Mitra S (2008). Long patch base excision repair in mammalian mitochondrial genomes. *J Biol Chem.* 283(39),26, 349-356.

4. Stuart J A, Hashiguchi K, Wilson D M, Copeland W C, Souza-Pinto N C and Bohr V A (2004). DNA base excision repair activities and pathway function in mitochondrial and cellular lysates from cells lacking mitochondrial *DNA Nucleic Acids Res.* 2004 32(7), 2181-2192.

5. Hashiguchi K, Bohr V A and de Souza-Pinto N C (2004).Oxidative stress and mitochondrial DNA repair: implications for NRTIs induced DNA damage. *Mitochondrion.* (2-3), 215-222.

6. Liu P and Demple B (2010). DNA repair in mammalian mitochondria: Much more than we thought? *Environ Mol Mutagen.* 51(5), 417-426.

Protocol No. 5.6: Determination of DNA Repair Synthesis Incorporation Activity in Mitochondria of Irradiated Cells and its Modulation by Radioprotective Drug

Assay Requirement

Cell culture plates, HEPES buffer (pH 7.6), EDTA, $MgCl_2$, BSA, KCl, DTT, phospho-creatine, creatine phosphokinase, ATP, dATP, dTTP, dGTP and 4 µM of dCTP, $\alpha^{32}P$-dCTP, glycerol, double-strand U-containing oligonucleotide (sequebnce mentioned below), heating oven, incubator, urea, Tris–borate/EDTA (TBE) buffer, acrylamide, bis-acrylamide, deionized water, phospho-Imager or equivalent scanner, gamma irradiator and cell culture facility.

5.6.1 Assay Procedure

5.6.1.1 Total DNA base excision repair assays involve incorporation of a radioactive nucleotide into a substrate containing a damage base (uracil-containing substrate can be used).

5.6.1.2 Cells or animals were divided in following four groups:

Gp 1: Control; untreated cells/animals

Gp 2: Irradiated cells/animals

Gp 3: Radioprotective Drug treated cells/animals

Gp 4: Irradiated cells/animals pretreated with radioprotective drug.

Followed by all treatments, mitochondria from treated cells or animal tissues were isolated using specific protocols mentioned above (refer Protocol No. 5.1).

5.6.1.3 The DNA repair (BER) reaction (10 µl) contained following components:

40 mM HEPES (pH 7.6)

0.1 mM EDTA

5 mM $MgCl_2$

0.2 mg/mL BSA

20 mM KCl

1 mM DTT

40 mM phosphocreatine

100 µg/mL creatine phosphokinase

2 mM ATP

40 µM of each dATP, dTTP, dGTP and 4 µM of dCTP

0.8 µCi α^{32}P-dCTP

3 per cent glycerol and

double-strand U-containing oligonucleotide

50-ATA TAC CGC GG(**U**) CGG CCG ATC AAG CTT ATT

30-TAT ATG GOG CC G GCC GGC TAG TTC GAA TAA

5.6.1.4 The reaction was started by the mixing of the cell lysates, equal to 10–50 µg protein per reaction.

5.6.1.5 The reactions mixture was kept at 37 °C for 3h and stopped by adding 2.5 µg of proteinase K and 0.5 µl of 10 per cent SDS and again incubating at 55 °C for 30 minutes.

5.6.1.6 The DNA was then precipitated overnight at -20 °C after addition of 1µg glycogen, 4 µl of 11 M ammonium acetate, 63 µl of ethanol.

5.6.1.7 Samples were centrifuged, dried and mixed with 10 µl of formamide loading dye and resolved using gel electrophoresis.

5.6.1.8. The whole DNS repair (BER) activity was calculated in terms of ^{32}P-dCTP signal intensity of the product band relative to a control samples. Background radioactivity should be subtracted from the reaction mixture without adding the protein lysate.

5.6.2 Electrophoresis

5.6.2.1 20 per cent polyacrylamide gels containing 7M urea, in 1x Tris–borate/EDTA (TBE) buffer was used.

5.6.2.2 The acrylamide solution was prepared by diluting a ready to use commercially available acrylamide/bis-acrylamide solution (37:1) in H$_2$O and adding 7M urea and 0.1 volume of 10x TBE.

5.6.2.3 Followed by urea solubilization, the solution was filtered.

5.6.2.4 That stock can be stored up to 2 months in the refrigerator, provided protected from the light.

5.6.2.5 1mm gels of at least 16 cm length should be used for better resolution.

5.6.2.6 The samples are loaded and the gels resolved in TBE buffer at 16W, for 1h.

5.6.2.7 The signals are then acquired using a Phospho-Imager or equivalent scanner.

5.6.2.8 The radioactivity in the lanes was quantified using the Image Quant software (1-10).

References

1.   Maynard S, de Souza-Pinto N, Scheibye-Knudsen M, Bohr V A (2010). Mitochondrial base excision repair assays. *Methods.* 51, 416–425.

2. Latimer J J and Kelly C M (2014). Unscheduled DNA synthesis: the clinical and functional assay for global genomic DNA nucleotide excision repair. *Methods Mole. Biol.* 1105, 511-532

3. Biggerstaff M and Wood R D (2006). Repair synthesis assay for nucleotide excision repair activity using fractionated cell extracts and UV-damaged plasmid DNA. *Methods Mol. Biol.* 314, 417-434.

6. Shen J C, Fox E J, Ahn E H and Loeb L A (2013). A rapid assay for measuring nucleotide excision repair by oligonucleotide retrieval. *Scientific Reports.* 4, 4894. DOI: 10.1038/Srep04894

7. Matsumoto M, Yaginuma K, Igarashi A, Imura M, Hasegawa M, Iwabuchi K, Date T, Mori T, Ishizaki K, Yamashita K, Inobe M and Matsunaga T (2007). Perturbed gap-filling synthesis in nucleotide excision repair causes histone H_2AX phosphorylation in human quiescent cells. *J. Cell Sci.* 120, 1104-1112.

8. Zou L and Elledge S J (2003). Sensing DNA damage through ATRIP recognition of RPA-ssDNA complexes. *Science.* 300(5625), 1542-1548.

9. Bradbury J M and Jackson S P (2003). The complex matter of DNA double-strand break detection. *Biochem. Soc. Trans.* 31, 40-44.

10. Szczesny B, Tann A W, Longley M J, Copeland W C and Mitra S (2008). Long patch base excision repair in mammalian mitochondrial genomes. *J. Biol. Chem.* 283(39), 26349-56.

11. Stuart J A, Hashiguchi K, Wilson D M, Copeland W C, Souza-Pinto N C and Bohr V A (2004). DNA base excision repair activities and pathway function in mitochondrial and cellular lysates from cells lacking mitochondrial DNA. *Nucleic Acids Res.* 32(7), 2181-2192.

12. Page M M and Stuart J A (2009). *In vitro* measurement of DNA base excision repair in isolated mitochondria. *Methods Mol Biol.* 554:213-231.

Protocol No. 5.7: Evaluation of Radioprotective Drug for its Gene-Specific mtDNA Repair Capabilities in Irradiated Cells/Mitochondria

Assay Requirement

Cell culture facility, methylene blue, restriction enzyme PvuII, agarose, Tris–HCl (pH 7.5), KCl, EDTA, Fpg protein, agarose, alkaline loading buffer, Ficoll, EDTA, bromocresol purple, NaOH, HindIII, southern Blot apparatus, electrophoresis unit, riboprobe, gamma irradiator and phospho-Imager.

5.7.1 Assay Procedure

5.7.1.1 Cells Cultured Using Standard Methods were Divided into Following Four Groups

Gp 1: Control; untreated cells

Gp 2: Irradiated cells

Gp 3: Radioprotective drug treated cells

Gp 4: Irradiated cells pretreated with radioprotective drug.

Followed by all treatments, cells were treated with methylene blue using protocol mentioned below:

5.7.1.2 Methylene Blue (MB) Treatment Assay

5.7.1.2.1 Cultured approximately~80 per cent confluence gamma irradiated cells were washed with PBS (plus calcium and magnesium). The concentration of methylene blue (MB), time of treatment and level of irradiation should be between 20 and 200 µM, 0.5 and 2 h, and 1-2 Gy respectively. The level of damage should be an average of 1–1.5 lesions/fragment.

5.7.1.3 DNA Isolation from the Treated Cells and Restriction Digestion

5.7.1.3.1 The DNA was isolated from the treated cells and digested with the help of restriction enzymes using standard protocols described elsewhere.

5.7.1.3.2 Restriction enzyme PvuII was useful to linearize human mtDNA.

5.7.1.3.3 Restriction efficiency should be checked quickly on 0.8 per cent agarose mini gel. The DNA must be fully restricted before further processing.

5.7.1.3.4 Purified DNA should be prepared and restricted as a positive control.

5.7.1.4 Treatment of DNA with Fpg

Note:

Serial dilution of Fpg was made in 1x reaction buffer (100 mM Tris–HCl pH 7.5–8.0, 100 mM KCl, 2 mM EDTA) at 0.01, 0.1, 1, 10 ng/µl and added to the sample tubes containing damaged and undamaged DNA. The optimum concentration of Fpg protein and reaction time should be optimized on the basis of optimal enzyme:DNA ratios, which can be determined by running optimization experiments with purified DNA.

*The maximum enzyme (Fpg): DNA ratio at which little or no cutting was seen in the control (undamaged) DNA after 20 minutes considered as highest enzyme: DNA ratio that can be used in the assay.

* The lowest enzyme; Fpg: DNA ratio and reaction time where specific incision i.e number of incisions observed in undamaged DNA was subtracted from number of incisions seen in damaged (MB-treated) DNA].

5.7.1.4.1 Added 2.2µg of DNA in 33µl of 1x reaction buffer for each sample.

5.7.1.4.2 15 µl of this reaction mixture added into following sets:

Set 1. Tube containing 5 µl of the pre-determined Fpg dilution,

Set 2. Tube containing 5 µl 1x reaction buffer.

5.7.1.4.3 Samples was placed in waterbath at 37°C and incubated for pre-determined time (optimize time 10, 20, 40 and 80 minutes).

5.7.1.4.4 The reactions was stopped at the appropriate times by the addition of 2.2 µl of 10x alkaline loading buffer [composition: 25 per cent Ficoll, 10 mM EDTA, 0.025 per cent (w/v) bromocresol purple, 0.5 M NaOH. Reaction mixture was mixed and incubated at 37°C for 15 minutes to complete denaturation of the DNA.

5.7.1.4.5 At the same time, an aliquot of HindIII-digested DNA molecular weight marker should be prepared similarly as positive control and denatured in the same way.

5.7.1.4.6 All the reaction mixture sets, i.e. control and tests samples were subjected to Southern blotting.

5.7.1.5 Alkaline Gel Electrophoresis and Southern Blot Analysis

5.7.1.5.1 For a 20 kb gene of interest 0.6 per cent (w/v) agarose gel should be used. For a linearized mammalian mitochondrial genome, a 0.75 per cent gel should be used.

5.7.1.5.2 For transfer standard southern Blot apparatus should be used.

5.7.1.5.3 The riboprobe against the gene of interest can be prepared using appropriate kit.

5.7.1.5.4 The use of a riboprobe was recommended because it allows strand-specific annealing if desired and effective stripping of the membrane.

5.7.1.5.5 The riboprobe template plasmids can be prepared by cloning the mitochondrial sequences amplified by PCR into the PCRII vector between the T7 and SP6 promoter sequences.

5.7.1.5.6 Quantification of the band intensities was estimated using Phosphor-imager due to its extended linear range (1-10).

5.7.1.5.7 The average number of incisions for each sample was calculated using the Poisson distribution using following formula:

Avarage no. of incision= Band intensity in a minus enzyme lane/band intensity in an enzyme-treated lane.

References

1. Bohr V A (1991). Assessment of DNA damage and repair in specific genome region by quantitative immune coupled PCR. *Carcinogenesis.* 2, 1983–1992.

2. Anson R M, Mason P A and Bohr V A (2006). Gene-specific and mitochondrial repair of oxidative DNA damage. *Methods Mol. Biol.* 314, 155–181.

3. Thorslund T, Sunesen M, Bohr V A and Stevnsner T (2002). Repair of 8-oxog is slower in endogeneous nuclear genes than in mitochondrial DNA and is without stranilbias, DNA Repair (Amst), (In: *Handbook of Models for Human Aging*, Ed: P. Michael Conn). 1, 261–273.

4. Bohr VA (1994). Gene specific damge and repair of DNA adducts and cross-links. *IARC Sci. Publ.* 125,361–369.

5. Bohr VA and Anson RM (1995). DNA damage, mutation and fine structure DNA repair in aging. *Mutat Res* 338:25–34.

6. Anderson S, Bankier AT, Barrell BG, de Bruijn M H, Coulson A R, Drouin J, Eperon I C, Nierlich D P, Roe B A, Sanger F, Schreier P H, Smith A J, Staden R and Young IG (1981). Sequence and organization of the human mitochondrial genome. *Nature.* 290, 457–465.

7. Santos J H, Meyer J N, Skorvaga M, Annab L A, Van H B (2004). Mitochondrial hTERT exacerbates free-radical-mediated mtDNA damage. *Aging Cell.* 3, 399–411.

8. Yala-Torres S, Chen Y, Svoboda T, Rosenblatt J and Van HB (2000). Role of mitochondrial DNA damage in the development of diabetic retinopathy and the metabolic memory phenomenon associated with its progression. *Methods.* 22, 135–147.

9. Van H B, Chen Y, Nicklas J A, Rainville I Rand O'Neill J P (1998). Development of long PCR techniques to analyze deletion mutations of the human hprt gene. *Mutat Res.* 403, 171–175.

10. Maynard S, de Souza-Pinto N, Scheibye-Knudsen M and Bohr V A (2010). Mitochondrial base excision repair assays. *Methods.* 51, 416–425.

Protocol No. 5.8: Immuno-fluorescent Detection of Gamma Radiation Induced 8-oxoG mtDNA Damage and Repair Using 8-oxoG-specific Antibodies

Note

Gamma radiation induced oxidative damage to DNA, that can be detected and quantify by measuring the concentration of oxidative DNA adduct i.e. 8-oxoG using immunofluorescence detection method. Immunofluorescence based detection of 8-oxoG adducts did not require mitochondrial isolation. It also permits to the investigators to analyze 8-oxoG DNA adduct localization in either nucleus or mitochondrial locations. This method also reveals the kinetics of oxidative lesion formation and repair at cellular level. Various research groups and commercial manufacturers develop monoclonal antibodies to detect 8-oxoG in nuclear or mitochondrial DNA.

Assay Requirement

Cell culture facility, slides with coverslip, H_2O_2 menadione, PBS, methanol, RNase, Triton X-100, 8-oxoG antibody, Fpg, Mut M, FLARE buffer, DNase, proteinase K, NaOH, ethanol, paraformaldehyde, HCl, goat serum, Alexa Fluor 488 anti-mouse secondary antibody, DAPI.

5.8.1 Assay Procedure

5.8.1.1 Cells cultured using standard methods were divided in following four groups:

Gp 1: Control; untreated cells

Gp 2: Cells treated with gamma irradiation/other DNA oxidizing agent

Gp 3: Cells treated with ant-iradiation drug

Gp 4: Irradiated/oxidizing agent treated cells pretreated with anti-radiation drug.

5.8.1.2 Cells grown on covered slides were treated with gamma radiation (2-4 Gy) or oxidizing agent, such as H_2O_2 (*e.g.* 500 µM for 30 minutes) or menadione (*e.g.* 50 µM for 30 minutes)/radioprotective drug, in serum free medium Medium was replaced with fresh complete medium and incubated for 0, -24h.

5.8.1.3 Cells were washed thrice with cold PBS. Slides were fixed with cold methanol for 20 minutes at -20 °C temperature.

5.8.1.3 Followed by washing twice with PBS at room temperature, the slides were treated with RNase (100 μg/ml) solution for 1h at 37 °C temperature. Complete digestion of RNA in the fixed cells is mendatory to get the clear immunofluorescent signal.

5.8.1.4 The cells can be washed for 3.0 minutes in 1 per cent Triton X-100 in PBS, for cellular permeabilization. The cells again need to wash and replace with normal PBS.

5.8.1.5 The cells on slides were incubated with 8-oxoG antibody. Slides were then incubated with Fpg 1 U/μl in FLARE buffer or Mut M (10 μg/ml) or DNase I (1 U/μl) for 1h at 37 °C temperature. The final 8-oxoG signal should be significantly reduced in these slides.

5.8.1.6 After washing with PBS, the slides were treated with proteinase K (10 μg/ml for 7 minutes (at room temperature to inhibit cross reactivity of the 8-oxoG antibody to proteins.

5.8.1.7 For immunodetection of 8-oxoG in mitochondrial DNA, the cells were treated for 5-7 minutes with NaOH (50 mM in 50 per cent ethanol).) and then refix the cells in 4 per cent paraformaldehyde in PBS at this step for 10 minutes at room temperature.

5.8.1.8 While, for immunodetection of nuclear/genomic DNA 8-oxoG signals, the slides were need to be washed with PBS and treated with 4N HCl for 5-7 minutes at room temepreture to denature the DNA.

5.8.1.9 Cells were blocked overnight with 10 per cent goat serum in PBS at 4°C temperature.

5.8.1.10 To detect the oxidized DNA adduct, 8-oxoG antibody treatment at dilution of 1:250 in 10 per cent goat serum-PBS for 1h at 37 °C and the Alexa Fluor 488 anti-mouse secondary antibody at the 1:1000 dilution for 1h at 37 °C in the dark was performed.

5.8.1.11 The cells on slides were then mounted with DAPI-containing Vectashield and analyzed with fluorescence microscope.

5.8.1.12 Negative controls consisted of secondary antibody-only and normal mouse IgG in place of the primary antibody were also prepared similarly and analysed with the similar fashion (1-10).

References

1. Ohno M, Oka S and Nakabeppu Y (2009). Quantitative analysis of oxidized guanine, 8 oxoguanine, in mitochondrial DNA by immunofluorescence method. *Methods Mol. Biol.* 554,199-212.

2. Soultanakis R P, Melamede R J, Bespalov I A, Wallace S S, Beckman K B, Ames B N, Taatjes D J andJanssen-Heininger Y M (2000).Fluorescence detection of 8-oxoguanine in nuclear and mitochondrial DNA of cultured cells using a recombinant Fab and confocal scanning laser microscopy. *Free Radic. Biol. Med.* 28(6), 987-98.

3. Wang Z, Rhee D B, Lu J, Bohr C T, Zhou F, Vallabhaneni H, de Souza-Pinto N C andLiu Y (2010). Characterization of oxidative guanine damage and repair in mammalian telomeres. *PLoS Genet.* 6(5), e1000951.

4. See comment in PubMed Commons belowHazra T K, Hill J W, Izumi T and Mitra S (2001). Multiple DNA glycosylases for repair of 8-oxoguanine and their potential *in vivo* functions. *Prog. Nucleic Acid Res. Mol. Biol.* 68, 193-205.

5. Oka S, Ohno M and Nakabeppu Y (2009). Construction and characterization of a cell line deficient in repair of mitochondrial, but not nuclear, oxidative DNA damage. *Methods Mol. Biol.* 554, 251-64.

6. Sheng Z, Oka S, Tsuchimoto D, Abolhassani N, Nomaru H, Sakumi K, Yamada H and Nakabeppu Y (2012). 8-Oxoguanine causes neurodegeneration during MUTYH-mediated DNA base excision repair. *J Clin Invest.* 122(12), 4344-4361.

7. Hazra T K, Izumi T, Maidt L, Floyd R A and Mitra S (1998).The presence of two distinct 8-oxoguanine repair enzymes in human cells: their potential complementary roles in preventing mutation. *Nucleic Acids Res.* 26(22), 5116-5122.

8. Maynard S, de Souza-Pinto N, Scheibye-Knudsen M and Bohr V A (2010). Mitochondrial base excision repair assays. *Methods.* 51, 416–425.

9. See comment in PubMed Commons belowBespalov I A, Bond J P, Purmal A A, Wallace S S and Melamede R J (1999). Fabs specific for 8-oxoguanine: control of DNA binding. *J. Mol. Biol.* 293(5), 1085-1095.

10. See comment in PubMed Commons belowPersinger R L, Melamede R, Bespalov I, Wallace S, Taatjes D J and Janssen-Heininger Y (2001). Imaging techniques used for the detection of 8-oxoguanine adducts and DNA repair proteins in cells and tissues. *Exp. Gerontol.* 36(9),1483-1494.

Protocol No. 5.9: *In vitro* Assays for the Determination of Histone Deacetylase Activity in Irradiated Cells/Animals and its Modulation by Radioprotective Compound Treatment using HPLC without Extraction of Decetylated Product

Introduction

Gamma radiation is known to induce deacetylation of DNA binding histone protein, by induction of histone deacetylases (HDACs) enzymes activity. Histone deacetylases contribute to the posttranslational modifications of histones as well as other non-histone proteins. As the results of deacetytation, histone protein conformation altered lead to functional impairment. Though role of histone deacetylayion is not very much understood, however, it is very clear that deacetylation of histone difinetly play an important role in radiosensitization/radioprotection. Gamma irradiation induced decaetylation of DNA bound histone and thus provides an excess to the radiation induced free radicals to DNA. Therefore, compounds having decatylase inhibitory activity can provide protection to the DNA and thus can developed as radioprotector.

Assay Requirement

Cell culture facility, Fluorescent histone deacetylase substrate MAL solution, 7-hydroxycoumarin, DMSO, ethanol, Histone deacetylase, NaH_2PO_4, Na_2HPO_4, EDTA, NaCl, glycerol, mercaptoethanol, HCl, sodium acetate, HPLC system with fluorescence detector and RP-18, 5mm column (size 250x4.6 mm), mobile phase: acetonitrile/water/trifluoro acetic acid (55:45:0.01v/v), deacetylase inhibitors in DMSO or methanol or ethanol or water, centrifuge and gamma irradiator.

5.9.1 Assay Procedure

5.9.1.1 To evaluates *in vivo* histone deacetylase inhibitory activities, animals/culture cells were divided into following four groups:

    i. Control untreated animals/cells

    ii. Animals/cells irradiated with gamma radiation

    iii. Animals/cells treated with histone deacetylase inhibitor

    iv. Irradiated animals pretreated with histone deacetylase inhibitor

The tissue homogenates of the animals organs preferably liver homogenate were prepared and total soluble proteins were extracted using standard procedures.

5.9.1.2 Following four experimental groups were formed:

Gp. 1: **Substrate with incubation buffer:** 10-12µl of the MAL (substrate) stock solution and10-15µl of the stock solution of 7-hydroxycoumarin was added with incubation buffer to a total volume of 1ml. [*For inhibitor (radioprotector) screening, compound was dissolved with incubation buffer to a concentration of 12-fold higher than the desired assay concentration*]

Gp.2: **Standards without enzyme activity (negative control):**110 µl of incubation buffer was added with 10µl of the substrate (MAL)/internal standard solution (7-hydroxycoumarin) and mix well on ice.

Gp.3: **Inhibitor compound with enzymes:** 100µl of enzyme (from commercial source) or from rate liver homogenate (dilute the enzyme to 100 µl with incubation buffer) was taken. 10 µl of diluted inhibitors (radioprotective compound under evaluation) in different concentrations was mixed with enzyme solution gently. Reaction mixture was kept on ice for 10 minutes. 10µl of the substrate (MAL)/ internal standard solution (7-hydroxycoumarin) was added and mixed gently. Reaction mixture was allowed to incubate at 4°C for another 15 minutes.

Gp.4: **Enzyme activity standards (positive control):**100µl of diluted enzyme (either from the commercial source or from the rate liver homogenate) was added with 10µl of the buffer. Reaction mixture was incubated on ice for 10 minutes. 10µl of the substrate/internal standard (7-hydroxycoumarin) solution was mixed to the reaction mixture and allowed to kept at 4°C for another 15 minutes.

5.9.1.3 All reaction mixtures and standards were incubated at 37 °C at least for 90-100 minutes.

5.9.1.4 1000µl of acetonitrile was added to all reaction mixtures with gentle shaking for 1 minute.

5.9.1.5 Reaction mixture was centrifuged at 10,000 rpm for 5 minutes.

5.9.1.6 600µl supernatant was collected and out of which 20 µl injected into the HPLC system. The system was run with fluorescence detector (excitation wavelength 330nm, emission wavelength 395 nm) flow rate 0.5ml/minute.

5.9.1.7 Chromatograph obtained was analyzed for the peak positions at retention time 5.06 minutes for the deacetylated substrate, 5.64 minutes for the internal standard (7-hydroxycoumarin), and 7.09 minutes for the substrate (MAL).

5.9.1.8 Quantitative estimation of the the deacetylated product and its inhibition was carried out using chromatographic quantification software (1-8).

References

1. Heltweg B, Trapp J and Jung M (2005). *In vitro* assays for the determination of histone deacetylase activity. *Methods* 36 332–337.

2. Heltweg B and Jung M (2002). A microplate reader-based nonisotopic histone deacetylase activity assay. *Anal Biochem.* 302(2),175-183.

3. Heltweg B and Jung M (2003). A homogeneous nonisotopic histone deacetylase activity assay. *J Biomol Screen.* 8(1). 89-95.

4. Hoffmann K, Brosch G, Loidl P and Jung M (1999). A non-isotopic assay for histone deacetylase activity. *Nucleic Acids Research,* 27(9), 2057–2058.

5. Zhang Y, LeRoy G, Seelig H P, Lane W S and Reinberg D (1998). The dermato-myositis-specific autoantigen Mi2 is a component of a complex containing histone deacetylase and nucleosome remodeling activities. *Cell.* 95(2), 279-289.

6. Guschin D, Wade P A, Kikyo N and Wolffe A P (2000). ATP-Dependent histone octamer mobilization and histone deacetylation mediated by the Mi-2 chromatin remodeling complex. *Biochemistry.* 39(18), 5238-5245.

7. Loidl P (1994). Histone acetylation: facts and questions. *Chromosoma.* 103(7), 441-449.

8. Drogaris P, Villeneuve V, Pomie's C, Lee E H, Bourdeau V, Bonneil E, Ferbeyre G, Verreault A and Thibault P (2011). Histone deacetylase inhibitors globally enhance h3/h4 tail acetylation without affecting H3 lysine 56 acetylation. *Scientific Reports.* 2, 220. DOI: 10.1038/srep00220.

Protocol No. 5.10: *In vitro* Determination of Gamma Radiation Induced Functional Activity Impairment of Histone Deacetylase and its Protection by Radioprotective Compounds using HPLC

Assay Requirement

Cell culture facility, Fluorescent histone deacetylase substrate MAL solution, 7-hydroxycoumarin, DMSO, ethanol, Histone deacetylase, NaH_2PO_4, Na_2HPO_4, EDTA, NaCl, glycerol, mercaptoethanol, HCl, sodium acetate, HPLC system with fluorescence detector and RP-18, 5mm column (size 250x4.6mm), mobile phase: acetonitrile/water/trifluoroacetic acid (55:45:0.01v/v), deacetylase inhibitors in DMSO or methanol or ethanol or water, centrifuge and gamma irradiator.

5.10.1 Assay Procedure

5.10.1.1 Following four experimental groups were formed:

Gp. 1: **Substrate with incubation buffer:** 10-12µl of the MAL (substrate) stock solution and 10-15µl of the stock solution of 7-hydroxycoumarin was added with incubation buffer to a total volume of 1ml. [*For inhibitor (radioprotector) screening, compound was dissolved with incubation buffer to a concentration of 12-fold higher than the desired assay concentration.*

Gp. 2: **Negative control (without enzyme):** 110µl incubation buffer was added with 10µl of the substrate (MAL)/internal standard solution (7-hydroxycoumarin). The mixture was mixed well and kept on ice.

Gp. 3: **Irradiation of Histone deacetylase:** 100µl of histone deacetylase (stock solution of 1mg/ml) was irradiated with different dose (3-7kGy) of gamma radiation in absence (Set A) and presence (Set B) of different concentrations of radioprotective compound under evaluation.

5.10.1.2 100µl of irradiated enzyme was mixed with 100 µl of incubation buffer and reaction mixture kept on ice for 10 minutes.

5.10.1.3 10µl of the substrate (MAL)/internal standard (7-hydroxycoumarin) solution added to the reaction mixture and allow to stand at 4 °C for another 15 minutes.

5.10.1.4 All samples and standards were incubated for at 37 °C, for at least 90-100 minutes.

5.10.1.5 Reaction was stopped by adding 1000µl of acetonitrile and vortex approximately 1 minute.

5.10.1.6 Reaction mixture was centrifuge at 10,000 rpm for 5 minutes and supernatant was collected

5.10.1.6 20µl of supernatant was injected into the HPLC system.The HPLC system was run with fluorescence detector (excitation wavelength 330 nm, emission wavelength 395 nm) flow rate 0.5ml/minutes.

5.10.1.7 Chromatograph obtained was analyzed for the peak positions appeared at retention time 5.06 minutes for deacetylated substrate, 5.64 minutes for the internal standard (7-hydroxycoumarin) and 7.09 minutes for the substrate (MAL).

5.10.1.8 Quantitative estimation was carried out using chromatographic quantification software (1-8).

References

1. Heltweg B, TrappJ and Jung M (2005). *In vitro* assays for the determination of histone deacetylase activity. *Methods*. 36, 332–337.

2. Heltweg B and Jung M (2002). A microplate reader-based nonisotopic histone deacetylase activity assay. *Anal Biochem*. 302(2),175-183.

3. Heltweg B and Jung M (2003). A homogeneous nonisotopic histone deacetylase activity assay. *J Biomol Screen*. 8(1). 89-95.

4. Hoffmann K, Brosch G, Loidl P and Jung M (1999). A non-isotopic assay for histone deacetylase activity. *Nucleic Acids Research*, 27(9), 2057–2058.

5. Zhang Y, LeRoy G, Seelig H P, Lane W S and Reinberg D (1998). The dermatomyositis-specific autoantigen Mi2 is a component of a complex containing histone deacetylase and nucleosome remodeling activities. *Cell*. 95(2), 279-289.

6. Guschin D, Wade P A, Kikyo N and Wolffe A P (2000). ATP-Dependent histone octamer mobilization and histone deacetylation mediated by the Mi-2 chromatin remodeling complex. *Biochemistry*. 39(18), 5238-5245.

7. Loidl P (1994). Histone acetylation: facts and questions. *Chromosoma*. 103(7), 441-449.

8. Drogaris P, Villeneuve V, Pomie's C, Lee E H, Bourdeau V, Bonneil E, Ferbeyre G, Verreault A and Thibault P (2011). Histone deacetylase inhibitors globally enhance h3/h4 tail acetylation without affecting H3 lysine 56 acetylation. *Scientific Reports*. 2, 220. DOI: 10.1038/srep00220.

Protocol No. 5.11: Determination of Histone Deacetylase Activity Modulation by Radioprotective Compounds using Fluorescence Spectrometric Analysis

Assay Requirement

Eosin Y, ethanol, fluorescence microplate reader, black 96-well microplates, reaction buffer [(0.5mM KH_2PO_4, 0.46mM NaOH (pH 8.0) was diluted with water in the ratio of 1: 59.4 ratio. Followed by this mixture was mixed with 39.6 parts of acetonitrile (buffer: water: acetonitrile 1: 59.4:39.6 ratio)], MAL substrate stock solution, histone deacetylase, HCl, ethyl acetate, fluorescence spectrophotometer, gamma irradiator.

5.11.1 Assay Procedure

5.11.1.1 To evaluate *in vivo* histone deacetylase inhibitory activities using fluorospectrometric analysis, animals/culture cells were divided into following four groups:

    i. Control untreated animals/cells

    ii. Animals/cells irradiated with gamma radiation

    iii. Animals/cells treated with histone deacetylase inhibitor

    iv. Irradiated animals cells pretreated with histone deacetylase inhibitor

The tissue homogenates of the animal organs preferably liver homogenate were prepared and total soluble proteins were extracted using standard procedures.

5.11.1.2 **Set 1:** 20µl of the MAL stock solution and 10µl of the stock solution of Eosin Y were mixed with incubation buffer to a total volume of 1 ml. Histone deacetylase inhibitor molecule should be diluted with incubation buffer.

5.11.1.3 **Set 2:** To analyze the standards without enzyme activity, 110µl of incubation buffer was added with 10µl of the substrate/internal standard solution and reaction mixture mixed gently. The reaction mixture was kept on ice.

5.11.1.4 **Set 3:** In another set of reaction mixture, 100µl of enzyme (stock 1mg/ml) was added with 10µl of diluted histone deacetylase inhibitor and mixed properly. Reaction mixture was allowed to incubate on ice for 10 minutes. Then 10µl of the substrate/internal standard solution was mixed with the reaction mixture and again incubated on ice for another 15 minutes.

5.11.1.5 **Set 4:** For enzyme activity analysis 100μl of enzyme or diluted enzyme was added with 10μl of incubation buffer. Reaction mixture was incubated on ice for 10 minutes. 10μl of the substrate/internal standard solution was added to the reaction mixture and kept on ice for another 15 minutes.

5.11.1.6 All the samples and standards were incubated at 37°C for 90-100 minutes.

5.11.1.7 Reaction was stopped by adding 400μl of 1M HCl and mixed well.

5.11.1.8 800μl of ethyl acetate was added to the reaction mixture and vortexed approximately for 1 minute.

5.11.1.9 Finally reaction mixture was centrifuged at 10,000rpm for 5 minutes to separate the supernatant.

5.11.1.10 400μl of supernatant was taken and upper phase containing ethyl acetate was evaporated using nitrogen purging. Evaporation can be done at room temperature without nitrogen just by incubation in the dark.

5.11.1.11 Residue obtained after evaporation was dissolved in 600μl of acetonitrile/buffer mixture, vortexes and transferred 250μl of it into microplate. Read the plate using Excitation wavelength 330nm and Emission wavelength 390nm (1-5).

References

1. Quinn A M and Simeonov A (2011). Methods for activity analysis of the proteins that regulate histone methylation. *Curr Chem Genomics.* 5(1), 95-105.

2. Baba R, Hori Y, Mizukami S and Kikuchi K (2012). Development of a fluorogenic probe with a transesterification switch for detection of histone deacetylase activity. *J. Am. Chem. Soc.* 134 (35), 14310–14313.

3. Wegener D, Hildmann C, Riester D and Schwienhorst A (2003). Improved fluorogenic histone deacetylase assay for high-throughput screening applications. *Anal. Biochem.* 321, 202-208.

4. Mazitschek R, Patel V, Wirth D F and Clardy J (2008). Development of a fluorescence polarization based assay for histone deacetylase ligand discovery. Bioorg. *Med. Chem. Lett.* 18, 2809–2812.

5. Fatkins D G and Zheng W (2008). A spectrophotometric assay for histone deacetylase. *Anal. Biochem.* 372, 82-88.

Section 6

Radiation Induced Mitochondrial Oxidative Phosphorylation Perturbations and their Modulation by Radioprotective Compounds Treatment

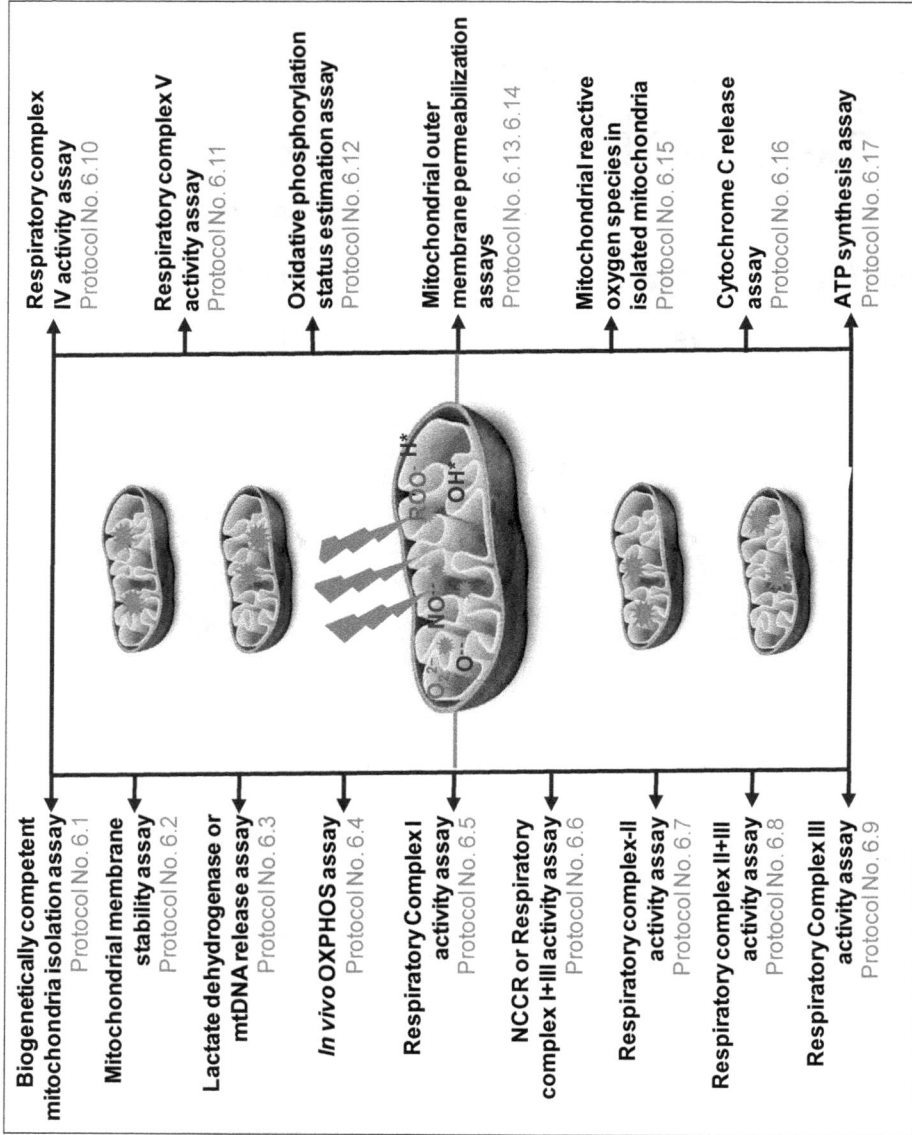

Figure 14: Schematic Summary of Assays Associated with Determination of Radiation Induced Mitochondrial Oxidative Phosphorylation Perturbations and their Protection by Radioprotective Drug Pretreatment.

Respiratory complex IV activity assay
Protocol No. 6.10

Respiratory complex V activity assay
Protocol No. 6.11

Oxidative phosphorylation status estimation assay
Protocol No. 6.12

Mitochondrial outer membrane permeabilization assays
Protocol No. 6.13, 6.14

Mitochondrial reactive oxygen species in isolated mitochondria
Protocol No. 6.15

Cytochrome C release assay
Protocol No. 6.16

ATP synthesis assay
Protocol No. 6.17

Biogenetically competent mitochondria isolation assay
Protocol No. 6.1

Mitochondrial membrane stability assay
Protocol No. 6.2

Lactate dehydrogenase or mtDNA release assay
Protocol No. 6.3

In vivo OXPHOS assay
Protocol No. 6.4

Respiratory Complex I activity assay
Protocol No. 6.5

NCCR or Respiratory complex I+III activity assay
Protocol No. 6.6

Respiratory complex-II activity assay
Protocol No. 6.7

Respiratory complex II+III activity assay
Protocol No. 6.8

Respiratory Complex III activity assay
Protocol No. 6.9

Protocol No. 6.1: Isolation of Biogenetically Competent Mitochondria from Mammalian Tissues and Cultured Cells to Study the Effect of Gamma Radiation

Introduction

Gamma radiation induced oxidative stress in cellular environment. Mitochondria are highly sensitive to oxidative stress because highest cellular oxidation takes place in mitochondria. Therefore, to study the mitochondrial oxidative stress subject to irradiation in presence and absence of a radioprotector, isolation of mitochondria is an essential step. The organelles (mitochondria) obtained using present protocol will be suitable for the investigation of biogenetic activities such as enzyme activity, mtDNA, mtRNA, mitochondrial protein synthesis and mitochondrial tRNA aminoacylation. In addition, mitochondria isolated by this process are capable to study protein import and mtDNA/protein interactions.

Assay Requirement

Sucrose, EDTA, Tris–HCl, sorbitol, EGTA, fatty acid-free bovine serum albumin (BSA), glass Potter–Elvehjem homogenizer with Teflon pestle, scissors, 50-ml tube, centrifuge, high-speed centrifuge, 1.5-ml Eppendorf tubes, ice bath, dextran T500 and Polyethylene Glycol 4000, 5mM potassium phosphate, EDTA, potassium phosphate, potassium EDTA, NaCl, KCl, $MgCl_2$, ADP, glutamate, malate, dNTP, $^{32}PdNTP$, $^{32}PUTP$, CTP, succinate, lysine and methionine.

6.1.1. Assay Procedure

6.1.1.1 Isolation of Mitochondria from Mammalian Tissue

6.1.1.1.1 Animals (male Wistar rats weighing 200–300 gm) or mice (20-25 gm) were killed by decapitation and the different organs removed, weighed, immediately placed on ice in the homogenization medium.

6.1.1.1.2 Homogenization medium composition: specific for liver and kidney (0.32M sucrose, 1mM EDTA, and 10mM Tris–HCl, pH 7.4)

or

Specific for brain and heart [(0.075 M sucrose, 0.225 M sorbitol, 1mM EGTA, 0.1 per cent fatty acid-free bovine serum albumin (BSA) and 10mM Tris–HCl, pH 7.4)].

6.1.1.1.3 Sterile solutions and glassware should be used. The tissues were washed to remove blood and connective tissue and cut into small pieces with a pair of scissors.

6.1.1.1.4 Fresh homogenization medium was added in a proportion of 4 ml/g of liver, 5 ml/g of kidney or brain, and 10 ml/g of heart.

6.1.1.1.5 Liver, kidney and heart homogenizations were performed in a glass Potter–Elvehjem homogenizer with a motor driven Teflon pestle with 4 strokes at 600 rpm.

6.1.1.1.6 The homogenization of brain was performed with 10–15 strokes in a Dounce-type glass homogenizer with a manually driven glass pestle.

6.1.1.1.7 Typically, a 250-gm rat has ~9 g liver, ~1.5 g brain, 1–1.5 g heart, and 2–3 g kidney (two).

6.1.1.1.8 In all cases, homogenates were transferred to a 50-ml tube and centrifuged for 5 minutes at 1000xg and 4 °C to pellet unbroken tissue, cells and nucleus.

6.1.1.1.9 The supernatants was then collected in a sterile 50-ml tube and kept on ice water. Since, many mitochondria can be entrapped in this first nuclear pellet (P1 fraction) the pellet resulting from first centrifugation with the remaining supernatant can be resuspended and rehomogenized again in fresh homogenization medium.

6.1.1.1.10 Then, a second round of centrifugation under the same conditions was performed. This is especially recommended when isolating mitochondria from the brain to increase the yield of organelles. If this additional step is carried out, the supernatant resulting from the second centrifugation (S-1b) is removed and gently mixed with the first one (S1a), discarding the nuclear pellet.

6.1.1.1.11 At this point the supernatants (S1) are centrifuged at 12,000xg and 4 °C for 10 minutes to pellet mitochondria (P2 fraction), resuspended in 10 vol of homogenization medium, and divided into 1.5 ml Eppendorf tubes.

6.1.1.1.12 This step can be bypassed when the source of organelles is liver or kidney. Transferring the S1 supernatant directly into eight to sixteen 1.5-ml Eppendorf tubes and discarding the remainder should yield a sufficient amount of mitochondria for most purposes.

6.1.1.1.13 This procedure is highly recommended when extracting mitochondria from liver. The Eppendorf tubes, each with 1.5 ml of S1 or resuspended P2, was centrifuged for 2 minutes in a microfuge tube at 13,000 rpm at 4°C.

6.1.1.1.14 Supernatant was discarded and the pellets that were in two Eppendorf tubes resuspended with 1.5 ml of homogenization medium and transferred to a single tube.

6.1.1.1.15 Tube was spun down again under the same conditions as before and the pellets from two Eppendorf tubes combined by resuspending them in 1.5 ml of homogenization medium. At the end of these two steps one obtained a single pellet that originated from four Eppendorf tubes.

6.1.1.1.16 The final pellets were washed by resuspension in 1ml of the appropriated incubation buffer, pelleted under similar conditions and again suspended in 1ml of the incubation buffer.

6.1.1.1.17 The resuspended mitochondria are maintained in a water–ice bath until used. The purpose of this series of washes in Eppendorf tubes was to free the mitochondrial fraction of particles containing high concentrations of harmful enzymes such as nucleases and proteases. This is particularly important in the isolation of liver and kidney mitochondria to analyze or purify mitochondrial RNAs.

Note: The mitochondrial pellet (P2) obtained in the high-speed centrifuge is too enriched in nucleases and is not suitable for in organello transcription assays

6.1.1.1.18 The yield of mitochondria measured as protein content can be estimated as one-third of the first supernatant for liver and all of the supernatant for the other tissues (1-4).

6.1.1.2 Mitochondria from Mammalian Cultured Cells

6.1.1.2.1 Mitochondria can be isolated from different cell lines by a modification of the above mentioned method as follows:

6.1.1.2.2 Exponentially growing cells can be harvested, washed two times with 1mM Tris–HCl (pH 7.0), 0.13M NaCl, 5mM KCl, 7.5mM $MgCl_2$, and broken in one half of the packed cell volume of 3.5mM Tris–HCl (pH 7.8), 2mM NaCl, 0.5mM $MgCl_2$, by using a Thomas homogenizer with a motor-driven Teflon pestle.

6.1.1.2.3 The homogenate should be mixed with one-ninth of the cell volume of 0.35M Tris–HCl (pH 7.8), 0.2M NaCl, 50mM $MgCl_2$, and spun for 3 minutes at 1600g to pellet unbroken cells, debris and nuclei.

6.1.1.2.4 The supernatant should be recentrifuged under the same conditions. The final supernatant is partitioned among Eppendorf tubes and spun at approximately 13,000xg in an Eppendorf microcentrifuge for 1 minute.

6.1.1.2.6 The mitochondrial pellets are washed serially with 35mM Tris–HCl (pH 7.8), 20mM NaCl, 5mM $MgCl_2$, then twice with 10mM Tris–HCl (pH 7.4), 1mM EDTA, 0.32M sucrose and finally resuspended in the appropriate incubation buffer. The entire purification process is performed at 4 °C (1-4).

6.1.1.3 Specific Incubation Requirements for different Analyses

6.1.1.3.1 In all types of assays, isolated organelles should be incubated in 0.5 ml of incubation buffer which contains 10mM Tris–HCl (pH 7.4), 25mM sucrose, 75mM sorbitol, 100mM KCl, 10mM K_2HPO_4, 0.05mM EDTA, 5mM $MgCl_2$, and 1 mg/ml BSA.

6.1.1.3.2 Incubation should be performed at 37 °C for the appropriated incubation time in a rotary shaker (12 rpm).

6.1.1.3.3 The smooth movement of the rotating wheel with the consequent inversion of the tubes allows good oxygenation of the mitochondrial suspension during incubation and prevents sedimentation of the organelles.

6.1.1.3.4 One should be aware that mitochondria can suffer irreversible damage and consequent impairment of their biogenetic activities if there is not appropriate oxygenation during the incubation period.

6.1.1.3.5 Although externally added ATP can support most of the biogenetic processes, it is recommended that the energy requirements of the mitochondria be provided by their own generated ATP.

6.1.1.3.6 To achieve this we usually add 1mM ADP, 10mM glutamate and 2.5mM malate to the incubation. A 50mM concentration of each dNTP(except the labeled one) and 20μCi of one $[^{32}P]$dNTP (800 Ci/mmol) are needed for in organelle mtDNA synthesis.

6.1.1.3.7 In addition, 10mM succinate instead of glutamate seems to yield better performance when using mitochondria isolated from kidney.

6.1.1.3.8 In the case of heart mitochondria a 10 minutes chase with cold UTP (0.2 mM) after the desired time of labeling improves the quality of the results significantly.

6.1.1.3.9 A mixture of all amino acids (except the labeled one) at final concentrations of 10–100μM each and 75μCi of the labeled amino acid, for example, ^{3H}lysine (77 Ci/mmol) or ^{35S}methionine (1200 Ci/mmol), can be supplied for aminoacylation or protein synthesis analyses (4-10).

References

1. Fernández-Vizarra E, López-Pérez M J and Enriquez J A (2002). Isolation of biogenetically competent mitochondria from mammalian. *Methods*. 26(4), 292-297.

2. Fernández-Vizarra E, Ferrín G, Pérez-Martos A, Fernández-Silva P, Zeviani M, and Enríquez J A (2010). Isolation of mitochondria for biogenetical studies: An update. *Mitochondrion*. 10(3), 253-262

3. Mercy L, Pauw A D, Payen L, Tejerina S, Houbion A, Demazy C, Raes M, Renard P, Arnould T (2005). Mitochondrial biogenesis in mtDNA-depleted cells involves a Ca^{2+} dependent pathway and a reduced mitochondrial protein import. *FEBS J*. 272(19), 5031-5055.

4. Wojtczak L and Kzab³ocki K (2008). Mitochondria in cell life, death and disease. *Postepy Biochem*. 54(2), 129-141.

5. Isolation of Mitochondria from Mammalian Tissues and Cultured Cells.Bio Encyclopedia.http://www.eplantscience.com/index/cell-biology-methods/isolation-of-mitochondria-from-mammalian-tissues-and-cultured-cells.php

6. Flierl A, Jackson C, Cottrell B, Murdock D, Seibel P and Wallace D C (2003). Targeted delivery of DNA to the mitochondrial compartment via import sequence-conjugated peptide nucleic acid. *Mol. Therapy*, 7, 550–557.

7. Barrientos A and Moraes C T (1998). Simultaneous transfer of mitochondrial DNA and single chromosomes in somatic cells: A novel approach for the study of defects in nuclear-mitochondrial communication. *Human Molecular Genetics*. 7(11),1801-1808.

8. Koulintchenko M, Temperley R J, Mason P A, Dietrich A and Lightowlers R N (2005). Natural competence of mammalian mitochondria allows the molecular investigation of mitochondrial gene expression. *Human Molecular Genetics*. 15(1),143-154.

9. Ekstrand M I, Falkenberg M, Rantanen A, Park C B, Gaspari M, Hultenby K, Rustin P, Gustafsson C M and Larsson N G (2004). Mitochondrial transcription factor A regulates mtDNA copy number in mammals. *Human Molecular Genetics*, 13 (9), 935–944.

10. Hansen A B, Griner N B, Anderson J P, Kujoth G C, Prolla T A, Loeb L A and Glick E (2006). Mitochondrial DNA integrity is not dependent on DNA polymerase-activity. *DNA Repair*, 571–579.

Protocol No. 6.2: Evaluation of Mitochondrial Membrane Stability in Irradiated Cells and its Protection by a Radioprotective Compound using Mito-Tracker Green and Mito-Tracker Red Probe

Assay Requirement

Cell culture facility, 48-well tissue culture plates, 35mm culture discs, Mito-Tracker Green, Mito-Tracker Red, fow cytometer and gamma irradiator.

6.2.1 Assay Procedure

6.2.1.1 Cells were grown using standard cell culture procedures.

6.2.1.2 Cells were plated for 24h in 48-well tissue culture plates/35mm culture discs in following four groups:

Set A: Control cells

Set B: Cells treated with radioprotective drug

Set C: Cells irradiated by gamma radiation

Set D: Irradiated cells pretreated with radioprotective drug

6.2.1.3 Followed by treatments cells were incubated with 200 nM Mito-Tracker Green or 200 nM Mito-Tracker Red for 1h.

6.2.1.4 Green and red fluorescence of the cells of all four groups were analyzed using a flow cytometer (1-7).

6.2.1.5 Mitochondrial membrane potential of the treated cells was represented as the ratio of fluorescence intensity of mitochondrial potential-sensitive Mito-Tracker Red (FL3 channel) and mitochondrial potential-independent Mito-Tracker Green (FL1 channel).

References

1. Cottet-Rousselle C, Ronot X, Leverve X and Mayol J F (2011). Cytometric assessment of mitochondria using fluorescent probes. 79A(6), 405–425.

2. Ramalho-Santos J, Amaral A, Sousa A P, Rodrigues A S, Martins L, Baptista M, Mota P C, Tavares R, Amaral S and Gamboa S (2007). *Modern Research and Educational Topics in Microscopy*. (Ed: A. Méndez-Vilas and J. Díaz). Formatex. pp. 394-402.

3. Krohn A J, Wahlbrink T and Prehn J H M (1999). Mitochondrial depolarization is not required for neuronal apoptosis. *The J. Neuroscience.* 19(17), 7394–7404.

4. http://www.mobitec.com/cms/products/bio/05-antibodies/apoptosis6. html? pdf=AZM7513.pdf.

5. http://tools.lifetechnologies.com/content/sfs/manuals/mp07510.pdf.

6. http://www.mobitec.de/probes/docs/sections/1202.pdf

7. http://www.nature.com/onc/journal/v29/n27/fig_tab/onc2010146ft.html

Protocol No. 6.3: *In vivo* Cell Death Determination Analysis by Analyzing Lactate Dehydrogenase, Hexosaminidase or Mitochondrial mtDNA Release in Plasma of Irradiated and Radioprotective Drug Treated Cells/Animals

Note

To analyze the occurrence of cell death in a particular pathology, one can detect in plasma the release by dying cells of certain factors such as LDH, hexosaminidase or mtDNA. MtDNA has been described in the plasma of trauma patients following extensive cell death caused by mechanical injury and was also detected in an experimental mouse model of systemic inflammatory response syndrome (SIRS) induced by TNF administration (1-7).

Assay Requirement

Blood vaccutainer, EDTA, p-nitrophenol-N-acetyl-beta-D-glucosaminide, primers specific for mitochondria-encoded genes (*i.e.* cytochrome b, cytochrome c oxidase III, and NADH oxidase).

6.3.1 Assay Procedure

6.3.1.1 Experimental animals were divided into following four groups:

   i. Control animals

   ii. Radioprotective Drug treated animals

   iii. Gamma radiation treated animals

   iv. Irradiated animals pretreated by radioprotective drug

6.3.1.2 Followed by animals treatment blood were collected from the orbital vein or tail (blood volume ~0.1 ml) at different time intervals, while for terminal measurement cardiac puncture was often used (blood volume ~1 ml).

6.3.1.3 Blood is sampled in tubes containing EDTA.

6.3.1.4 Cells from the blood were removed by centrifugation (10 minutes at 1000xg at 4°C). Additional centrifugation for 15 minutes at 2000xg was performed to deplete platelets.

6.3.1.5 The resulting supernatant was called plasma that can be used for measuring LDH or hexosaminidase or for detecting mtDNA.

6.3.1.6 For the hexosaminidase assay p-nitrophenol-N-acetyl-beta-D-glucosaminide sunstrate was used (3).

6.3.1.7 To purify the DNA from Blood ready to use DNA purification kit can be used.

6.3.1.8 Purified DNA was analyzed for mtDNA by quantitative PCR (1-7) using primers specific for three mitochondria-encoded genes (cytochrome b, cytochrome c oxidase III, NADH oxidase).

References

1. Berghe T V, Grootjans S, Goossens V, Dondelinger Y, Krysko D V, Takahashi N and Vandenabeele P (2013). Determination of apoptotic and necrotic cell death *in vitro* and *in vivo*. *Methods*. 61, 117–129.

2. Larsen T (2005). Determination of lactate dehydrogenase (LDH) activity in milk by a fluorometric assay. *J Dairy Res*. 72(2), 209-216.

3. Chagunda M G, Larsen T, Bjerring M and Ingvartsen K L (2006). L-lactate dehydrogenase and N-acetyl-beta-D-glucosaminidase activities in bovine milk as indicators of non-specific mastitis. *J Dairy Res*. 73(4), 431-440.

4. Augoff K and Grabowski K (2004). Significance of lactate dehydrogenase measurements in diagnosis of malignancies. *Pol. Merkur. Lekarski*. 17(102), 644-647.

5. Cerdán S, Rodrigues T B, Sierra A, Benito M, Fonseca L L, Fonseca C P and García-Martín M L (2006). The redox switch/redox coupling hypothesis. *Neurochem. Int*. 48(6-7), 523-530.

6. Ramírez B G, Rodrigues T B, Violante I R, Cruz F, Fonseca L L, Ballesteros P, Castro M M, García-Martín M L and Cerdán S (2007). Kinetic properties of the redox switch/redox coupling mechanism as determined in primary cultures of cortical neurons and astrocytes from rat brain. *J Neurosci Res*. 85(15), 3244-3253.

7. www.worthington-biochem.com/ldh/assay.html

Protocol No. 6.4: Determination of Gamma Radiation Effect on Oxidative Phosphorylation (OXPHOS) in Irradiated Cells and its Modulation by Radioprotective Drug/Compound: *In vitro* Assessment

Assay Requirement

Cell culture facility, respiration buffer, mannitol, KCl, $MgCl_2$, and K_2PO_4, KCN, polarograph with an adjustable chamber volume (*i.e.*, from 0.25 up to 1 ml), digitonin, bovine serum albumin, ADP, ethanol, toluidine blue, toluene blue, pyruvate, malate, glutamate, rotenone, KCN, succinate, ATP, malonate, glycerol 3-phosphate, G3P-dehydrogenase, FAD, duroquinol, antimycin A, ascorbate, tetramethyl-p phenylenediamine (TMPD), centrifuge, polarograph with an adjustable chamber volume (*i.e.*, from 0.25 up to 1 ml) and gamma irradiator.

6.4.1 Assay Procedure

6.4.1.1 Intact Cell-Coupled Endogenous Respiration

6.4.1.1.1 Collect exponentially growing cells (in medium changed a few hours before) by trypsinization and divided in following four groups:

Set A: Control cells (untreated)

Set B: Radioprotective drug treated cells

Set C: Cells irradiated by gamma radiation

Set D: Irradiated cells pretreated with radioprotective drug

Followed by drug and radiation treatment, cells were allowed to grow for 6-12 h with 5 per cent CO_2 concentration and 95 per cent humidity.

6.4.1.1.2 After incubation pellet the cells by contribution at 2500rpm.

6.4.1.1.3 Re-suspend the pellet at 5×10^6 cells/ml in respiration buffer (RB) consisting of 0.3M mannitol, 10mM KCl, 5 mM $MgCl_2$, and 10mM K_2PO_4, (pH 7.4).

6.4.1.1.4 Inject 300µl of cell suspension into the polarographic chamber and measure the intact cell coupled endogenous cell respiration.

6.4.1.1.5 Add KCN (to 700µM) to block respiration.

6.4.1.2 Digitonin Permeabilization of the Cells

6.4.1.2.1 Cell membranes can be permeabilized to different substrates by incubation with digitonin.

6 .4.1.2.2 The permeabilization can be done in the polarograph chamber. However, to better control the degree of permeabilization, it is suggested to do it outside the chamber immediately before the assay.

6.4.1.2.3 Add 60µg digitonin/ml to the cell suspension in RB (5×10^6 cells/ml) and incubate for 1–2 min at room temperature.

6.4.1.2.4 Add 5 vol RB supplemented with 1 mg/ml bovine serum albumin (BSA) to the cell suspension to stop desolubilization of the membranes.

6.4.1.2.5 Resuspend the pellet at 5×10^6 cells/ml RB supplemented with 1 mg/ml BSA and 0.5mM ADP.

6.4.1.2.6 Maintain the RB air-equilibrated at 37°C.

6.4.1.2.7 The suspension of permeabilized cells is now ready to use in oxygen consumption experiments.

Note

(*i*) *BSA and ADP should be freshly added to RB.*

(*ii*) *Eliminating impurities from the digitonin solution is advisable to obtain an appropriate and reproducible permeabilization of the cell membrane. Weight the appropriate quantity of digitonin, taking into account the degree of purity (e.g., Aldrich has a 75 per cent pure digitonin). Recrystallize the detergent by adding ethanol at a ratio of 40–50 mg digitonin/ml ethanol. Centrifuge the mixture 10s at 14,000g. Part of the impurities will be soluble in ethanol and should be discarded with the supernatant. Repeat the washing-out of the impurities three times. Air-dry the digitonin crystals. Resuspend the crystals in the appropriate volume of water. Still, the solution will not be totally clear. Do not heat it. Pass it through a blue tip several times and spin for 1 minutes at 14,000xg. The relatively small pellet obtained should be discarded since it represents a bulk of impurities. Use the supernatant as a pure digitonin solution.*

(*iii*) *The time of incubation with digitonin depends on the cell type and a time/ percentage of cells permeabilized curve should be obtained for each one.*

(*iv*) *After the permeabilization, take a drop of the cell suspension and add another drop of 0.5 per cent of toluidine blue. Check the degree of permeabilization by toluene blue exclusion under optic microscopy. About 95 per cent of the cells should be permeabilized (i.e., the dye accesses the cytoplasm and shows up blue). Avoid over permeabilization (1-7).*

6.4.1.3 Substrate Oxidation in Permeabilized Cells

6.4.1.3.1 Immediately after permeabilization, 300µl of cell suspension was injected into polarographic chamber. Oxygen consumption was started with the different substrates. If necessary, additional ADP can be added to the reaction mixture. Three classic experiments can be done (Figure 14):

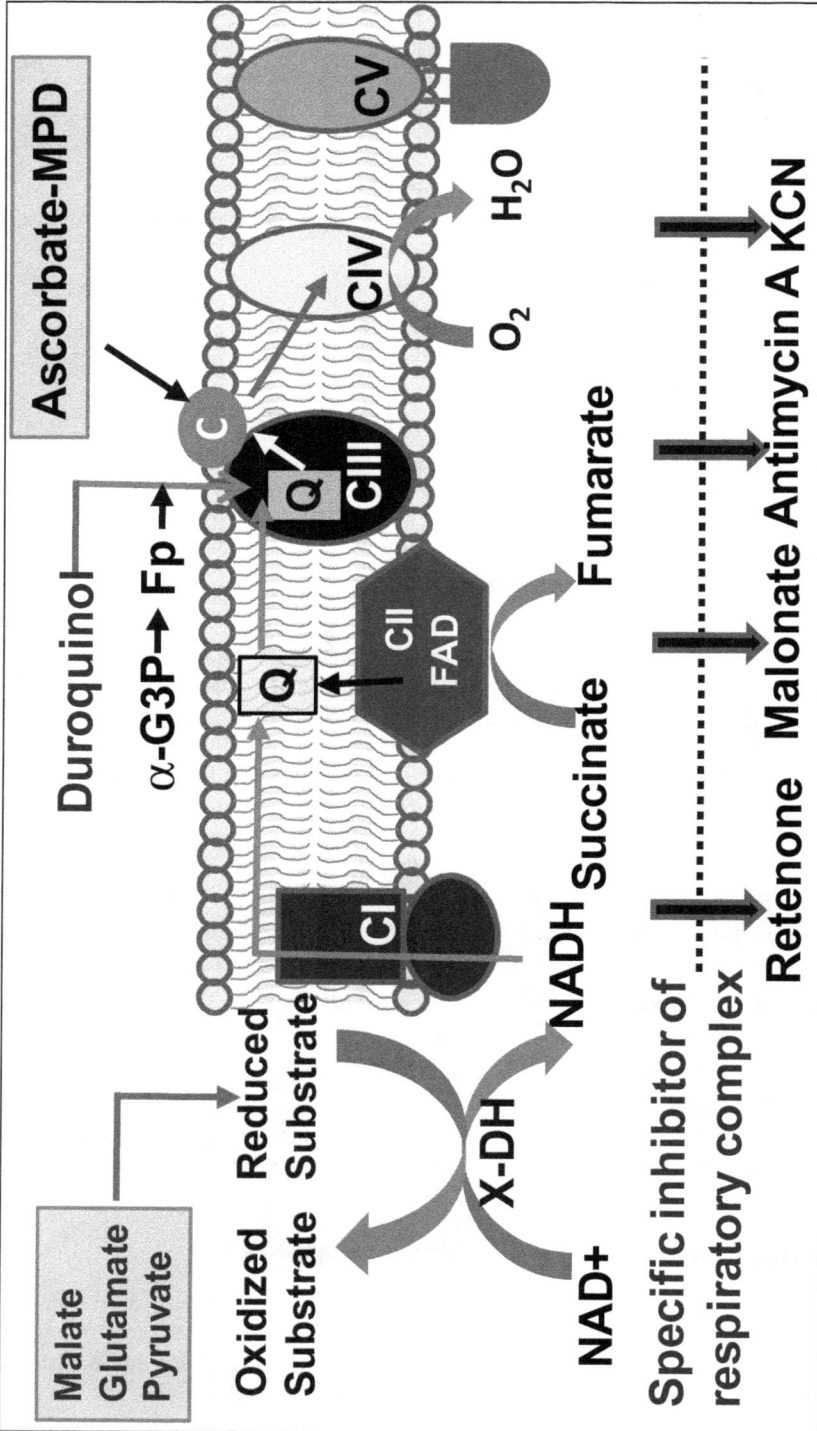

Figure 14: Schematic representation of the OXPHOS system showing the substrates and inhibitors used in the polarographic experiments. The lines indicate electron flow direction. C, cytochrome c; Fp, flavoprotein; X-DH, different specific dehydrogenases, a-G3P, glycerol 3-phosphate.

(i) Oxidation of pyruvate (8mM) plus malate (0.2 mM), followed by the oxidation of glutamate (15mM) (site I substrates), can be measured. The reaction is inhibited with rotenone (3lM) or KCN (700lM).

(ii) Oxidation of succinate (10mM) (site II substrate) in the presence of rotenone (3lM) and ATP (130lM) can be measured. After inhibition of complex II with malonate (10 mM), the oxidation of the glycerol 3-phosphate (G3P) (20mM) catalyzed by a G3P-dehydrogenase containing FAD (an enzyme associated with the mitochondrial internal membrane giving electrons to complex III) can be measured.

(iii) Oxygen uptake triggered by duroquinol (0.6mM) and inhibited with antimycin A (1µM), which measures the activity of complex III plus IV, was monitored subsequently.

6.4.1.3.2 In another experiment, the oxidation of ascorbate (10mM) plus N; N; N0; N0- tetramethyl-p phenylenediamine (TMPD, 0.2 mM), through complex IV, allows an estimation of complex IV activity (Figure 14), which is inhibited with 700lM KCN (7-15).

References

1. Barrientos A (2002). *In vivo* and in organello assessment of OXPHOS activities. *Methods* 26 307–316.

2. Chretien D, Enit P B, Chol M, Lebon S, Rotig A, Munnich A and Rustin P (2003). Assay of mitochondrial respiratory chain complex I in human lymphocytes and cultured skin fibroblasts. *Biochem. Biophys. Res. Commun.* 30(1), 222–224.

3. Triepels R H, Van Den Heuvel L P, Trijbels J M and Smeitink J A (2001). Respiratory chain deficiency. *Am. J. Hum. Genet.* 106, 37–45.

4. Barrientos A, Fontanesi F and Díaz F (2009). Evaluation of the mitochondrial respiratory chain and oxidative phosphorylation system using polarography and spectrophotometric enzyme assays. *Curr Protoc Hum Genet.* doi: 10.1002/0471142905.hg1903s63.

5. Acin-Perez R, Fernandez-Silva P, Peleato M L, Perez-Martos A and Enriquez J A (2008). Respiratory active mitochondrial super-complexes. *Mol. Cell.* 32, 529–539.

6. Birch-Machin M A and Turnbull D M (200). Assaying mitochondrial respiratory complex activity in mitochondria isolated from human cells and tissues. *Methods Cell. Biol.*, 65:97–117

7. Fernandez-Vizarra E, Tiranti V and Zeviani M (2009). Assembly of the oxidative phosphorylation system in humans: what we have learned by studying its defects. *Biochim. Biophys. Acta.* 1793, 200–211.

8. Medja F, Allouche S, Frachon P, Jardel C, Malgat M, de Camaret B M, Slama A, Lunardi J, Mazat J P and Lombès A (2009). Development and implementation of standardized respiratory chain spectrophotometric assays for clinical diagnosis. *Mitochondrion.* 9(5), 331-339. doi: 10.1016/j.mito.2009.05.001.

9. Ragan C I and Cottingham I R (1985). The kinetics of quinone pools in electron transport. *Biochim. Biophys. Acta*. 811,13–31.

10. Villani G, Greco M, Papa S, Attardi G (1998). Low reserve of cytochrome C oxidase capacity *in vivo* in the respiratory chain of a variety of human cell types. *J. Biol. Chem.* 273, 31829–31836.

11. Medvedev E S, Couch V A and Stuchebrukhov A A (2010). Determination of the intrinsic redox potentials of FeS centers of respiratory complex I from experimental titration curves. *Biochim Biophys Acta*. 1797(9), 1665–1671.

12. Cogliati S, Frezza C, Soriano M E, Varanita T, Quintana-Cabrera R, Corrado M, Cipolat S, Costa V, Casarin A, Gomes L C, Perales-Clemente E, Salviati L, Fernandez-Silva P, Enriquez J A and Scorrano L (2013). Mitochondrial cristae shape determines respiratory chain supercomplexes assembly and respiratory efficiency. *Cell*, 155 (1), 160–171.

13. Medvedev E S, Couch V A and Stuchebrukhov A A (2010). Determination of the intrinsic redox potentials of FeS centers of respiratory complex I from experimental titration curves. *Biochimica. Biophysica. Acta*. 1797(9), 1665–1671.

14. Schägger H and Pfeiffer K (2001). The ratio of oxidative phosphorylation complexes I–V in bovine heart mitochondria and the composition of respiratory chain supercomplexes. *J. Biol. Chem.* 276, 37861-37867.

15. Stroh A, Anderka O, Pfeiffer K, Yagi T, Finel M, Ludwig B and Schägger H (2004). Assembly of respiratory complexes I, III, and IV into NADH oxidase supercomplex stabilizes complex I in *Paracoccus denitrificans*. *J. Biol. Chem.* 279, 5000-5007.

Protocol No. 6.5: Determination of Gamma Radiation Effect on Oxidative Phosphorylation (OXPHOS) in Isolated Mitochondria of Irradiated Cells and its Modulation by Radioprotective Drug/Compound: Complex I Activity Estimation

Assay Requirement

Cell culture facility, trypsin, phosphate-buffered saline, -80 °C defreezer, NADH, Tris, BSA, KCN, antimycin A, 2,3-dimethoxy-5- methyl-6-n-decyl-1,4-benzoquinone (DB), rotenone, spectrophotometer and gamma irradiator.

6.5.1 Assay Procedure

Harvested exponentially growing cells by trypsinizion and divided in following four groups:

Set A: Control cells (untreated)

Set B: Radioprotective drug treated cells

Set C: Cells irradiated by gamma radiation

Set D: Irradiated cells pretreated with radioprotective drug

Followed by drug and radiation treatment, cells were allowed to grow for 6-12 h with 5 per cent CO_2 concentration and 95 per cent humidity.

6.5.1.1 Exponentially growing cells are further collected by trypsinization, pelleted, and resuspended in cold phosphate-buffered saline medium to a concentration of 5×10^6 cells/ml.

6.5.1.2 The suspension was aliquoted and frozen at -80 °C until used for the different enzymatic assays.

6.5.1.3 The activity of the five OXPHOS enzymes can be measured in isolated mitochondria (refer Protocol No. 6.1)

6.5.1.4 A total of 20–40μg of mitochondrial protein will be sufficient to determine the activity of each complex.

6.5.1.5 The assays described here are performed at 37°C (except the citrate synthase, at 30°C) in 1ml of medium (1-6).

(A) Simplified representation of complex I. Only the extrinsic and intrinsic arms have been represented although the complex is composed of more than 40 polypeptides. The electron transfer pathway is represented by the lines and arrows. FMN, flavin mononucleotide.

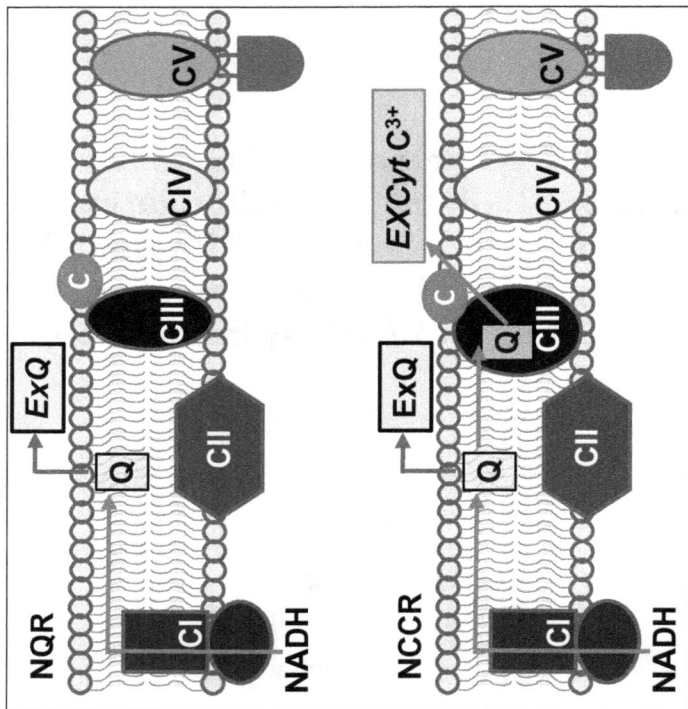

(B) Measurement of the rotenone-sensitive NADH decylubiquinone oxidoreductase (NQR) and NADHcytochrome c oxidoreductase (NCCR) activities. Initial electron donors are underlined and final electron acceptors are in italics. EXQ, exogenous quinone; EXCyt C3p, exogenous oxidized cytochrome C.

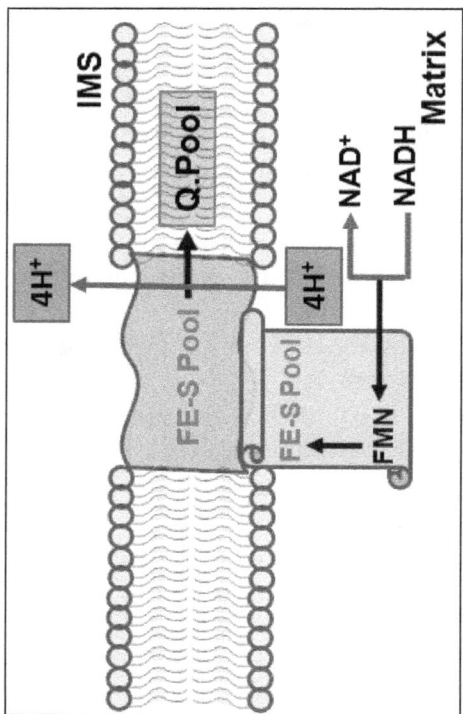

Figure 15a,b: Complex I or NADH-ubiquinone Oxidoreductase.

6.5.2 Complex-I. Measurement of the Rotenone-Sensitive NADH Decylubiquinone Oxidoreductase Activity (NQR)

6.5.2.1 The assay is performed at 340nm following the decrease in absorbance resulting from the oxidation of NADH (Figure 15B).

6.5.2.2 In 800µl of H_2O was added into isolated mitochondria (20–40µg of protein, see notes for pretreatment of mitochondria) and incubate 1–2 minutes at 37 °C.

6.5.2.3 Added 200µl of 50mM Tris (pH 8.0) medium supplemented with 5 mg/ml BSA, 0.8 mM NADH as donor, 240µM KCN, and 4µM antimycin A (AA).

6.5.2.4 The reaction was started by the addition of 50µM of the acceptor 2,3-dimethoxy-5- methyl-6-N-decyl-1,4-benzoquinone (DB).

6.5.2.5 The activity was measured for 3 minutes Addition of 4µM rotenone and measurement of the activity for 3 additional minutes allowed quantification of the rotenone-sensitive activity spectrophotometrically at 340nm (4-12).

Note

(i) *It is not possible to measure NQR in whole cells due to the absence of activation of the enzyme with decylubiquinone.*

(ii) *Because the accuracy of the method relies on the access of NADH to its binding site in complex I, mitochondrial membranes should be disrupted first by freeze–thawing the samples two or three times in hypotonic medium (25mM K_2PO_4 pH 7.2, 5mM $MgCl_2$) followed by a hypotonic shock in H_2O. The incubation of mitochondria in H_2O must not be longer than three minutes, because the rotenone-sensitive activity can be affected.*

(iii) *The rate of rotenone inhibition depends on the quality of the mitochondrial preparation and it is more accurate if BSA is present in the medium.*

(iv) *NADH should be freshly prepared. It is suggested that its spectra should be run to ensure that it is not oxidized.*

(v) *Considerable rotenone-insensitive NQR activity in fibroblast mitochondrial fractions has been reported, and washing the mitochondria in hypotonic buffer appears to partially remove the activity of the contaminating enzyme NADH: Cytochrome b5 reductase resulting in an increase in the rotenone sensitive component of NQR activity (1-15).*

References

1. Barrientos A (2002). *In vivo* and in organello assessment of OXPHOS activities. *Methods*. 26 307–316.

2. Chretien D, Enit P B, Chol M, Lebon S, Rotig A, Munnich A and Rustin P (2003). Assay of mitochondrial respiratory chain complex I in human lymphocytes and cultured skin fibroblasts. *Biochem. Biophys. Res. Commun.* 30(1), 222–224.

3. Triepels R H, Van Den Heuvel L P, Trijbels J M and Smeitink J A (2001). Respiratory chain deficiency. *Am. J. Hum. Genet.* 106, 37–45.

4. Barrientos A, Fontanesi F and Díaz F (2009). Evaluation of the mitochondrial respiratory chain and oxidative phosphorylation system using polarography and spectrophotometric enzyme assays. *Curr. Protoc. Hum. Genet.* doi:10.1002/0471142905.hg1903s63.

5. Acin-Perez R, Bayona-Bafaluy M P, Fernandez-Silva P, Moreno-Loshuertos R, Perez-Martos A, Bruno C, Moraes C T and Enriquez J A (2004). Respiratory complex III is required to maintain complex I in mammalian mitochondria. *Mol. Cell.* 13, 805–815.

6. Acin-Perez R, Fernandez-Silva P, Peleato M L, Perez-Martos A and Enriquez J A (2008). Respiratory active mitochondrial super-complexes. *Mol. Cell.* 32, 529–539.

7. Birch-Machin M A and Turnbull D M (200). Assaying mitochondrial respiratory complex activity in mitochondria isolated from human cells and tissues. *Methods Cell. Biol.*, 65:97–117.

8. Chretien D, Rustin P, Bourgeron T, Rotig A, Saudubray J M and Munnich A (1994). Reference charts for respiratory chain activities in human tissues. *Clin. Chim. Acta.* 228, 53–70.

9. Fernandez-Vizarra E, Tiranti V and Zeviani M (2009). Assembly of the oxidative phosphorylation system in humans: what we have learned by studying its defects. *Biochim. Biophys. Acta.* 1793, 200–211.

10. Ragan C I and Cottingham I R (1985). The kinetics of quinone pools in electron transport. *Biochim. Biophys. Acta.* 811, 13–31.

11. Medvedev E S, CouchV A and. Stuchebrukhov A A (2010). Determination of the intrinsic redox potentials of FeS centers of respiratory complex I from experimental titration curves. *Biochim. Biophys. Acta.* 1797(9), 1665–1671.

12. Cogliati S, Frezza C, Soriano M E, Varanita T, Quintana-Cabrera R, Corrado M, Cipolat S, Costa V, Casarin A, Gomes L C, Perales-Clemente E, Salviati L, Fernandez-Silva P, Enriquez J A and Scorrano L (2013). Mitochondrial cristae shape determines respiratory chain supercomplexes assembly and respiratory efficiency. *Cell*, 155 (1), 160–171.

13. Medvedev E S, Couch V A and Stuchebrukhov A A (2010). Determination of the intrinsic redox potentials of FeS centers of respiratory complex I from experimental titration curves. *Biochimica. Biophysica. Acta.* 1797(9), 1665–1671.

14. SchäggerH and PfeifferK (2001).The ratio of oxidative phosphorylation complexes I–V in bovine heart mitochondria and the composition of respiratory chain supercomplexes. *J. Biol. Chem.* 276, 37861-37867.

15. Stroh A, Anderka O, Pfeiffer K, Yag iT, Fine M, Ludwig B and Schägger H (2004). Assembly of respiratory complexes I, III, and IV into NADH oxidase supercomplex stabilizes complex I in *Paracoccus denitrificans*. *J. Biol. Chem.* 279, 5000-5007.

Protocol No. 6.6: Determination of Gamma Radiation Effect on Oxidative Phosphorylation (OXPHOS) in Isolated Mitochondria of Irradiated and its Modulation by Radioprotective Drug/Compound: Measurement of the Rotenone-Sensitive NADH Cytochrome C Oxidoreductase Activity (NCCR) or Complex I+III Activity

Assay Requirement

Cell culture facility, Tris, BSA, oxidized cytochrome C, KCN, NADH, rotenone, spectrophotometer and gamma irradiator.

6.6.1 Assay Procedure

6.5.1.1 Collect exponentially growing cells (in medium changed a few hours before) by trypsinization and divided in following four groups:

Set A: Control cells (untreated)

Set B: Drug treated cells

Set C: Cells irradiated by gamma radiation

Set D: Irradiated cells pretreated with radioprotective drug

Followed by drug and radiation treatment, cells were allowed to grow for 6-12 h with 5 per cent CO_2 concentration and 95 per cent humidity.

6.6.2 Enzymatic Activity Estimation

6.4.2.1 Exponentially growing cells (in medium changed a few hours before) are collected by trypsinization, pelleted, and resuspended in cold phosphate-buffered saline medium to a concentration of 5×106 cells/ml.

6.5.2.2 The suspension was aliquoted and frozen at -80 °C until used for the different enzymatic assays.

6.5.2.3 The activity of the five OXPHOS enzymes can be measured in isolated mitochondria (refer Protocol No. 6.1).

6.5.2.4 A total of 20–40µg of mitochondrial protein will be sufficient to determine the activity of complex I to III.

6.5.2.5 The assays described here are performed at 37°C (except the citrate synthase, at 30°C) in 1ml of medium.

6.6.3 Complex I to III Activity Estimation

6.5.3.1 The assay was performed at 340nm following the increase in ab resulting from the reduction of cytochrome c (Figure 15b).

6.5.3.2 800µl of H_2O was added to isolated mitochondria (20–40µg of p and incubated for 1–2 minutes at 37 °C.

6.5.3.3 Added 200µl of 50mM Tris (pH 8.0) medium supplemented with 5 ml BSA, 40µM oxidized cytochrome C as the acceptor and 240µM I

6.5.3.4 After 4 minutes of incubation, start the reaction by addition of 0.8 NADH as donor.

6.5.3.5 Measure activity for 3 minutes.

6.5.3.6 The addition of 4µM rotenone and measurement of activity for thre additional minutes allow quantification of the rotenone sensitive activity spectrophometrically at 340nm (1-14).

References

1. Barrientos A (2002). *In vivo* and in organello assessment of OXPHOS activities. *Methods* 26 307–316.

2. Chretien D, Enit P B, Chol M, Lebon S, Rotig A, Munnich A and Rustin P (2003). Assay of mitochondrial respiratory chain complex I in human lymphocytes and cultured skin fibroblasts. *Biochem. Biophys. Res. Commun.* 30(1), 222–224.

3. Barrientos A, Fontanesi F and Díaz F (2009). Evaluation of the mitochondrial respiratory chain and oxidative phosphorylation system using polarography and spectrophotometric enzyme assays. *Curr. Protoc. Hum. Genet.* doi:10.1002/0471142905.hg1903s63.

4. Acin-Perez R, Bayona-Bafaluy M P, Fernandez-Silva P, Moreno-Loshuertos R, Perez-Martos A, Bruno C, Moraes CT and Enriquez J A (2004). Respiratory complex III is required to maintain complex I in mammalian mitochondria. *Mol. Cell.* 13, 805–815.

5. Acin-Perez R, Fernandez-Silva P, Peleato M L, Perez-Martos A and Enriquez J A (2008). Respiratory active mitochondrial super-complexes. *Mol. Cell.* 32, 529–539.

6. Chretien D, Slama A, Briere JJ, Munnich A, Rotig A and Rustin P (2004). Revisiting pitfalls, problems and tentative solutions for assaying mitochondrial respiratory chain complex III in human samples. *Curr. Med. Chem.* 11, 233–239.

7. Fernandez-Vizarra E, Tiranti V and Zeviani M (2009). Assembly of the oxidative phosphorylation system in humans: what we have learned by studying its defects. *Biochim. Biophys. Acta.* 1793, 200–211.

8. Krahenbuhl S, Talos C, Wiesmann U and Hoppel CL (1994). Development and evaluation of a spectrophotometric assay for complex III in isolated mitochondria, tissues and fibroblasts from rats and humans. *Clin. Chim. Acta.* 230, 177–187.

9. Medja F, Allouche S, Frachon P, Jardel C, Malgat M, de Camaret B M, Slama A, Lunardi J, Mazat J P and Lombès A (2009). Development and implementation of standardized respiratory chain spectrophotometric assays for clinical diagnosis. *Mitochondrion*. 9(5), 331-339. Doi: 10.1016/j.mito.2009.05.001.

10. Ragan C I and Cottingham I R (1985). The kinetics of quinone pools in electron transport. *Biochim. Biophys. Acta*. 811, 13–31.

11. Medvedev E S, Couch V A and. Stuchebrukhov A A (2010). Determination of the intrinsic redox potentials of FeS centers of respiratory complex I from experimental titration curves. *Biochim. Biophys. Acta*. 1797(9), 1665–1671.

12. Slipetz D M, Aprille J R, Goodyer P R and Rozen R (1991). Deficiency of complex III of the mitochondrial respiratory chain in a patient with facioscapulohumeral disease. *Am. J. Hum. Genet*. 48(3), 502–510.

13. Schägger H and Pfeiffer K (2001). The ratio of oxidative phosphorylation complexes I–V in bovine heart mitochondria and the composition of Respiratory chain supercomplexes. *J. Biol. Chem*. 276, 37861-37867.

14. Stroh A, Anderka O, Pfeiffer K, Yagi T, Finel M, Ludwig B and Schägger H (2004). Assembly of respiratory complexes I, III, and IV into NADH oxidase supercomplex stabilizes complex I in *Paracoccus denitrificans*. *J. Biol. Chem*. 279, 5000-5007.

Protocol No. 6.7: Determination of Gamma Radiation Effect on Oxidative Phosphorylation (OXPHOS) in Isolated Mitochondria of Irradiated Cells and its Modulation by Radioprotective Drug/ Compound: Measurement of Complex-II Activity

Introduction

Succcinate: Ubiquinone reductase or complex II consists of four nuclear DNA-encoded polypeptides (Figure 16a). Two of them form the peripheral part of the enzyme and act as the enzyme succinate dehydrogenase in the tricarboxylic acid cycle. They are associated with the membrane through two integral "anchor" proteins. Electrons coming from the oxidation of succinate to fumarate are channeled through complex II to ubiquinone. In complex II, a flavin is linked covalently to a 70-kDa peptide, the flavoprotein subunit (Fp), which is associated with the iron–protein subunit (Ip) containing three nonheme Fe–S centers and form the succinate dehydrogenase or SDH. One of the anchor proteins seems to contain cytochrome, b which would be involved with electron transport to ubiquinone (Figure 16a).

Assay Requirement

Cell culture facility, 2,6-dichloro-phenolindophenol, KH_2PO_4, EDTA, BSA, DCPIP, rotenone, ATP, succinate, decylubiquinone, malonate, succinate, spectrophotometer, gamma irradiator.

6.7.1 Assay Procedure

6.7.1.1 Exponentially growing cells was harvested (in medium changed a few hours before) by trypsinization and divided in following four groups:

Set A: Control cells (untreated)

Set B: Radioprotective drug treated cells

Set C: Cells irradiated by gamma radiation

Set D: Irradiated cells pretreated with radioprotective drug

Followed by drug and radiation treatment, cells were allowed to grow for 6-12 h with 5 per cent CO_2 concentration and 95 per cent humidity.

6.7.2 Enzymatic Activity Estimation

6.7.2.1 Exponentially growing cells (in medium changed a few hours before) are collected by trypsinization, pelleted and resuspended in cold phosphate-buffered saline medium to a concentration of 5×10^6 cells/ml.

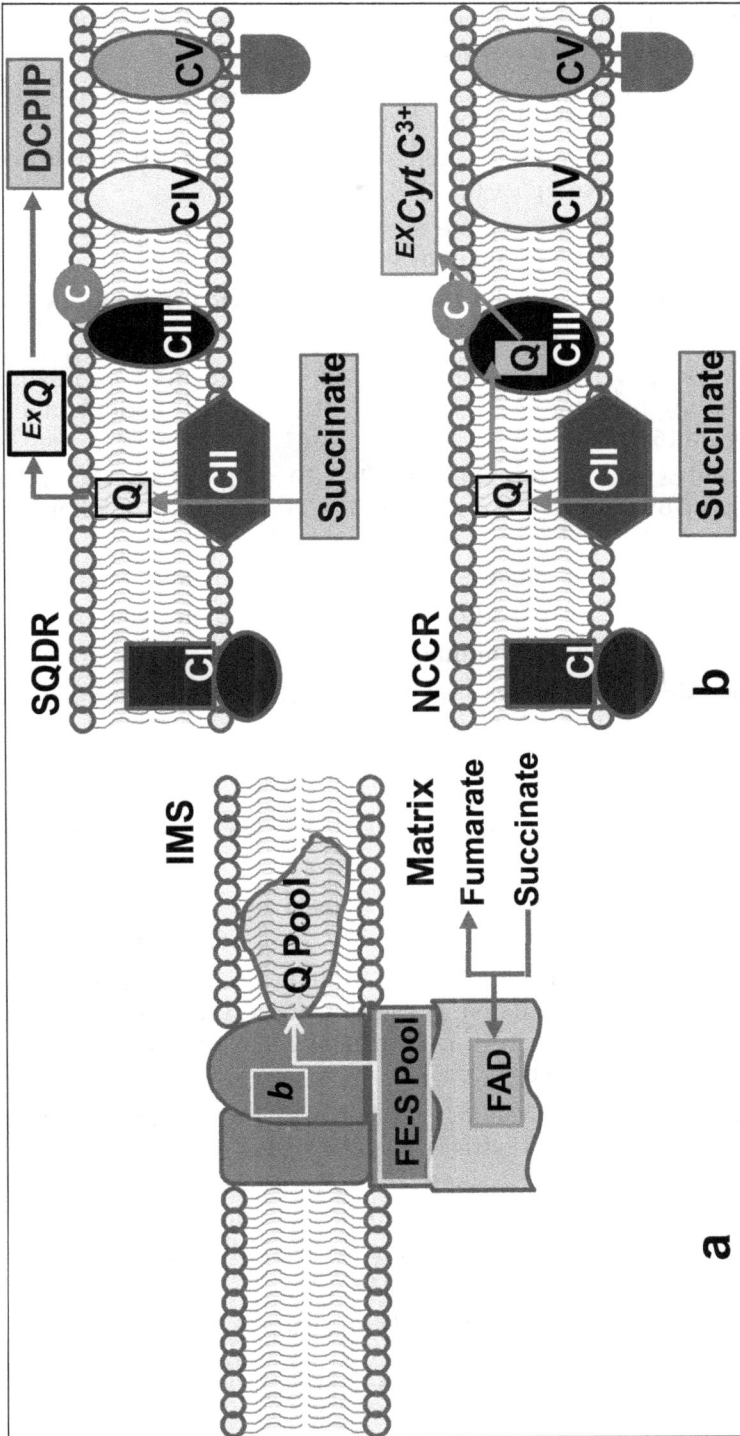

Figure 16: (a) Simplified Representation of Complex-II Molecular Arrangement; (b) Schematic Representation of Succinate Decylubiquinone DCPIP Reductase (SQDR) Activity; Excyt C* Exogenous Oxidized Cytochrome C.

6.7.2.2 The suspension was aliquoted and frozen at -80 °C until used for the different enzymatic assays.

6.7.2.3 The activity of the five OXPHOS enzymes can be measured in isolated mitochondria (refer Protocol No. 6.1).

6.7.2.4 A total of 20–40µg of mitochondrial protein will be sufficient to determine the activity of complex-II.

6.7.2.5 The assays described here are performed at 37°C (except the citrate synthase, at 30°C) in 1ml of medium.

6.7.3 Measurement of Succinate Decylubiquinone DCPIP Reductase Activity (SQR)

6.7.3.1 The assay is performed at 600nm following the decrease in absorbance resulting from the reduction of 2,6-dichlorophenolindophenol (DCPIP; Figure 16B).

6.7.3.2 In 1ml of medium containing KH_2PO_4 10mM (pH 7.8), EDTA 2mM and 1 mg/ml BSA added mitochondria (20–40µg of protein) or whole cells (3–10 x 10^4 cells), 80µM DCPIP as acceptor, 4µM rotenone and 0.2mMATP.

6.7.3.3 Added 10mM succinate as the donor and incubate 10 minutes at 30 °C.

6.7.3.4 Start the reaction by addition of 80µM decylubiquinone.

6.7.3.5 Measure activity for 5 minutes.

6.7.3.6 The addition of 10mM malonate (competitive inhibitor) inhibits the oxidation of succinate (1-13).

References

1. Barrientos A (2002). *In vivo* and in organello assessment of OXPHOS activities. *Methods* 26 307–316.

2. Triepels R H, Van Den Heuvel L P, Trijbels J M and Smeitink J A (2001). Respiratory chain deficiency. *Am. J. Hum. Genet.* 106, 37–45.

3. Barrientos A, Fontanesi F and Díaz F (2009). Evaluation of the mitochondrial respiratory chain and oxidative phosphorylation system using polarography and spectrophotometric enzyme assays. *Curr Protoc Hum Genet.* doi:10.1002/0471142905.hg1903s63.

4. Birch-Machin M A and Turnbull D M (200). Assaying mitochondrial respiratory complex activity in mitochondria isolated from human cells and tissues. *Methods Cell. Biol.*, 65:97–117

5. Fernandez-Vizarra E, Tiranti V and Zeviani M (2009). Assembly of the oxidative phosphorylation system in humans: what we have learned by studying its defects. *Biochim. Biophys. Acta.* 1793, 200–211.

6. Krahenbuhl S, Talos C, Wiesmann U and Hoppel CL (1994). Development and evaluation of a spectrophotometric assay for complex III in isolated mitochondria, tissues and fibroblasts from rats and humans. *Clin. Chim. Acta.* 230,177–187.

7. Medja F, Allouche S, Frachon P, Jardel C, Malgat M, de Camaret B M, Slama A, Lunardi J, Mazat J P and Lombès A (2009). Development and implementation of standardized respiratory chain spectrophotometric assays for clinical diagnosis. *Mitochondrion.* 9(5), 331-339. doi: 10.1016/j.mito.2009.05.001.

8. Villani G, Greco M, Papa S and Attardi G (1998). Low reserve of cytochrome C oxidase capacity *in vivo* in the respiratory chain of a variety of human cell types. *J. Biol. Chem.* 273, 31829–31836.

9. Cogliati S, Frezza C, Soriano M E, Varanita T, Quintana-Cabrera R, Corrado M, Cipolat S, Costa V, Casarin A, Gomes L C, Perales-Clemente E, Salviati L, Fernandez-Silva P, Enriquez J A and Scorrano L (2013). Mitochondrial cristae shape determines respiratory chain supercomplexes assembly and respiratory efficiency. *Cell.* 155 (1), 160–171.

10. Slipetz D M, Aprille J R, Goodyer P R and Rozen R (1991). Deficiency of complex III of the mitochondrial respiratory chain in a patient with facioscapulohumeral disease. *Am. J. Hum. Genet.* 48(3), 502–510.

11. Schägger H and Pfeiffer K (2001). The ratio of oxidative phosphorylation complexes I–V in bovine heart mitochondria and the composition of Respiratory chain supercomplexes. *J. Biol. Chem.* 276, 37861-37867.

12. Stroh A, Anderka O, Pfeiffer K, Yagi T, Finel M, Ludwig B and Schägger H (2004). Assembly of respiratory complexes I, III, and IV into NADH oxidase supercomplex stabilizes complex I in *Paracoccus denitrificans. J. Biol. Chem.* 279, 5000-5007.

13. Acŷin-Pérez R, Fernández-Silva M P, Moreno-Loshuertos P, Pérez-Martos R, Bruno A, Moraes C, and Enrŷìquez J A (2004). Respiratory complex III is required to maintain complex I in mammalian mitochondria. *Mol. Cell.* 13(6), 805–815.

Protocol No. 6.8: Determination of Gamma Radiation Effect on Oxidative Phosphorylation (OXPHOS) in Isolated Mitochondria of Irradiated Cells and its Modulation by Radioprotective Drug/ Compound: Measurement of Succinate Cytochrome C Reductase (SCCR) or Complex II + III Activity

Assay Requirement

Cell culture facility, PBS, 2,6-dichloro-phenolindophenol, KH_2PO_4, EDTA, BSA, KCN, rotenone, rotenone, ATP, malonate, succinate, cytochrome C, spectrophotometer, gamma irradiator.

6.8.1 Assay Procedure

6.8.1.1 Collect exponentially growing cells by trypsinization and divided in following four groups:

Set A: Control cells (untreated)

Set B: Radioprotective drug treated cells

Set C: Cells irradiated by gamma radiation

Set D: Irradiated cells pretreated with radioprotective drug

Followed by drug and radiation treatment, cells were allowed to grow for 6-12 h with 5 per cent CO_2 concentration and 95 per cent humidity.

6.8.2 Enzymatic Activity Estimation

6.8.2.1 Exponentially growing cells (in medium changed a few hours before) are collected by trypsinization, pelleted, and resuspended in cold phosphate-buffered saline medium to a concentration of 5×10^6 cells/ml.

6.8.2.2 The suspension was aliquoted and frozen at −80 °C until used for the different enzymatic assays.

6.8.2.3 The activity of the five OXPHOS enzymes can be measured in isolated mitochondria.

6.8.2.4 A total of 20–40µg of mitochondrial protein will be sufficient to determine the activity of complex II and complex III.

6.8.2.5 The assays described here are performed at 37°C (except the citrate synthase, at 30°C) in 1ml of medium.

6.8.3 Measurement of Succinate Cytochrome C Reductase (SCCR) or Complex II + III

6.8.3.1 The assay was performed at 550nm following the increase in absorbance resulting from the reduction of cytochrome C.

6.8.3.2 In1ml of medium containing 10mMKH_2PO_4 (pH 7.8), 2mMEDTA and 1 mg/ml BSA add mitochondria (20–40µg of protein) or whole cells (3–10 x 10^4cells),240µM KCN, 4µM rotenone, and 0.2mM ATP.

6.8.3.3 Add 10mM succinate as the donor and incubate 10 minutes at 30 °C.

6.8.3.4 Start the reaction by addition of 40µM oxidized cytochrome C.

6.8.3.5 Measure activity for 5 minutes.

6.8.3.6 The addition of 10mMmalonate inhibits the oxidation of succinate (1-9).

References

1. Barrientos A (2002). *In vivo* and in organello assessment of OXPHOS activities. *Methods* 26 307–316.

2. Triepels R H, Van Den Heuvel L P, Trijbels J M and Smeitink J A (2001). Respiratory chain deficiency. *Am. J. Hum. Genet.* 106, 37–45.

3. Barrientos A, Fontanesi F and Díaz F (2009). Evaluation of the mitochondrial respiratory chain and oxidative phosphorylation system using polarography and spectrophotometric enzyme assays. *Curr Protoc Hum Genet.* doi:10.1002/0471142905.hg1903s63.

4. Krahenbuhl S, Talos C, Wiesmann U and Hoppel CL (1994). Development and evaluation of a spectrophotometric assay for complex III in isolated mitochondria, tissues and fibroblasts from rats and humans. *Clin. Chim. Acta.* 230, 177–187.

5. Huang L, Borders TM, Shen JT, Wang CJ, and Berry E (2005). Crystallization of mitochondrial respiratory complex II from chicken heart: A membrane protein complex diffracting to 2.0 Å. *Acta Crystallogr D Biol Crystallogr.* 61(Pt 4), 380–387.

6. Slipetz D M, Aprille J R, Goodyer P R and Rozen R (1991). Deficiency of complex III of the mitochondrial respiratory chain in a patient with facioscapulohumeral disease. *Am. J. Hum. Genet.* 48(3), 502–510.

7. Schägger H and Pfeiffer K (2001). The ratio of oxidative phosphorylation complexes I–V in bovine heart mitochondria and the composition of Respiratory chain supercomplexes. *J. Biol. Chem.* 276, 37861-37867.

8. Stroh A, Anderka O, Pfeiffer K, Yagi T, Finel M, Ludwig B and Schägger H (2004). Assembly of respiratory complexes I, III, and IV into NADH oxidase supercomplex stabilizes complex I in *Paracoccus denitrificans*. *J. Biol. Chem.* 279, 5000-5007.

9. Acŷin-Pérez R, Fernández-Silva M P, Moreno-Loshuertos P, Pérez-Martos R, Bruno A, Moraes C, and Enrŷìquez J A (2004). Respiratory complex III is required to maintain complex I in mammalian mitochondria. *Mol. Cell.* 13(6), 805–815.

Protocol No. 6.9: Determination of Gamma Radiation Effect on Oxidative Phosphorylation (OXPHOS) in Isolated Mitochondria of Irradiated Cells and its Modulation by Radioprotective Drug: Measurement of Complex III Enzyme Activity

Complex III Enzyme

Complex III, bc1 complex, or ubiquinol:cytochrome-c, oxidoreductase, is formed by three catalytic subunits—the mitochondrial DNA-encoded cyt b, cyt c1, and Rieske Fe–S protein and eight noncatalytic subunits. The enzyme catalyzes the transfer of two electrons from ubiquinol to ferry-cytochrome C and uses the free energy change to transport protons across the inner membrane (two from ubiquinol and two from the mitochondrial matrix), which contribute to the electrochemical gradient that drives aerobic synthesis of ATP. Complex III works through a modified Q-cycle mechanism depicted at Figure 17A.

Assay Requirement

Cell culture facility, KH_2PO_4, EDTA, BSA, decylubiquinol, KCN, rotenone, ATP, oxidized cytochrome C, antimycin A, reduced quinine, spectrophometer and gamma irradiator.

6.9.1 Assay Procedure

6.9.1.1 Collect exponentially growing cells (in medium changed a few hours before) by trypsinization and divided in following four groups:

Set A: Control cells (untreated)

Set B: Radioprotective drug treated cells

Set C: Cells irradiated by gamma radiation

Set D: Irradiated cells pretreated with radioprotective drug

Followed by drug and radiation treatment, cells were allowed to grow for 6-12 h with 5 per cent CO_2 concentration and 95 per cent humidity.

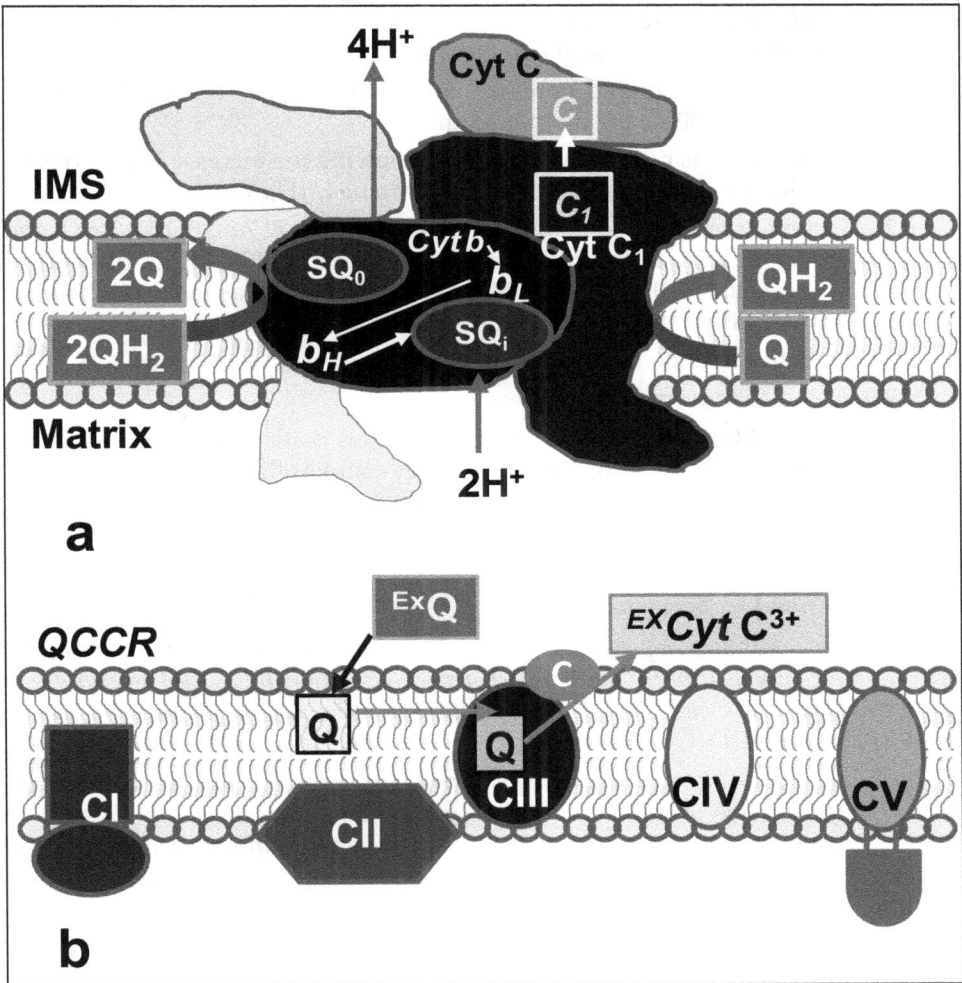

Figure 17: Complex III or Ubiquinol Cytochrome C Oxido-reductase.

(a), Schematic representation of complex III modified. (b), Measurement of ubiquinol cytochrome C reductase (QCCR). EXQ, exogenous quinone; EXCyt c3, exogenous oxidized cytochrome C.

6.9.2 Measurement of Complex III Enzyme Activity

6.9.2.1 Exponentially growing cells (in medium changed a few hours before) are collected by trypsinization, pelleted and resuspended in cold phosphate-buffered saline medium to a concentration of 5×10^6 cells/ml.

6.9.2.2 The suspension was aliquoted and frozen at -80 °C until used for the different enzymatic assays.

6.9.2.3 The activity of the five OXPHOS enzymes can be measured in isolated mitochondria.

6.9.2.4 A total of 20–40μg of mitochondrial protein will be sufficient to determine the activity of complex II and complex III.

6.9.2.5 The assays described here are performed at 37°C (except the citrate synthase, at 30°C) in 1ml of medium.

6.9.2.6 The assay was performed at 550nm following the increase in absorbance resulting from the reduction of cytochrome C (Figure 17).

6.9.2.7 In 1 ml of medium containing 10mM KH_2PO_4 (pH 7.8), EDTA 2mM, and 1 mg/ml BSA added to isolated mitochondria (20-40μg of protein) or whole cells (3-10 x 10^4 cells), 80mM decylubiquinol as the donor, 240μM KCN, 4μM rotenone, and 200μM ATP.

6.9.2.8 Start the reaction by addition of 40μM oxidized cytochrome C.

6.9.2.9 Measure activity for 5 minutes.

6.9.2.10 The addition of 0.4μM antimycin A allows distinguishing between the reductions of cytochrome C catalyzed by complex III and the nonenzymatic reduction of cytochrome C by the reduced quinone (1-11).

Note

Precuation need to be taken

(i) *Biological material should be freeze–thawed in isotonic medium before using, to gently disrupt the inner mitochondrial membrane. It has been described that addition of 0.6mM lauryl maltoside helps to disrupt the membranes and liberate the maximum QCCR activity.*

(ii) *Decylubiquinol must be freshly prepared. To the decylubiquinone solution adds a few crystals of lithium borohydrate and pipet it up and down until the solution becomes transparent. Eliminate the excess of borohydrate with a few microliters of concentrated HCl, until no bubbles are produced. The final pH should be 2–3.*

(iii) *Rotenone is used in the assay to inhibit complex I activity and prevent unspecific changes in the concentration of ubiquinol.*

(iv) *KCN inhibits complex IV activity preventing reoxidation of the reduced cytochrome c.*

(v) *The components of the reaction produce an unspecific oxidation of decylubiquinol, and also the inhibitors antimycin A (AA) affects the redox state of decylubiquinol. It is necessary, then, to run a control without biological material to obtain a background rate (with and without inhibitors of complex-III) that will be subtracted from the rate obtained with biological material. The background activity depends on the length and the composition of the side chain of the quinol used. Decylubiquinol is the most suitable substrate because of its low background activity (1-11).*

References

1. Barrientos A (2002). *In vivo* and in organello assessment of OXPHOS activities. *Methods* 26 307–316.

2. Barrientos A, Fontanesi F and Díaz F (2009). Evaluation of the mitochondrial respiratory chain and oxidative phosphorylation system using polarography and spectrophotometric enzyme assays. *Curr Protoc Hum Genet.* doi:10.1002/0471142905.hg1903s63.

3. Acin-Perez R, Bayona-Bafaluy M P, Fernandez-Silva P, Moreno-Loshuertos R, Perez-Martos A, Bruno C, Moraes CT and Enriquez J A (2004). Respiratory complex III is required to maintain complex I in mammalian mitochondria. *Mol. Cell.* 13, 805–815.

4. Birch-Machin M A and Turnbull D M (200). Assaying mitochondrial respiratory complex activity in mitochondria isolated from human cells and tissues. *Methods Cell. Biol.,* 65:97–117

5. Chretien D, Slama A, Briere J J, Munnich A, Rotig A and Rustin P (2004). Revisiting pitfalls, problems and tentative solutions for assaying mitochondrial respiratory chain complex III in human samples. *Curr. Med. Chem.*11, 233–239.

6. Krahenbuhl S, Talos C, Wiesmann U and Hoppel C L (1994). Development and evaluation of a spectrophotometric assay for complex III in isolated mitochondria, tissues and fibroblasts from rats and humans. *Clin. Chim. Acta.* 230, 177–187.

7. Medja F, Allouche S, Frachon P, Jardel C, Malgat M, de Camaret B M, Slama A, Lunardi J, Mazat J P and Lombès A (2009). Development and implementation of standardized respiratory chain spectrophotometric assays for clinical diagnosis. *Mitochondrion.* 9(5), 331-339. doi: 10.1016/j.mito.2009.05.001.

8. Slipetz D M, Aprille J R, Goodyer P R and Rozen R (1991). Deficiency of complex III of the mitochondrial respiratory chain in a patient with facioscapulohumeral disease. *Am. J. Hum. Genet.* 48(3), 502–510.

9. Schägger H and Pfeiffer K (2001). The ratio of oxidative phosphorylation complexes I–V in bovine heart mitochondria and the composition of Respiratory chain supercomplexes. *J. Biol. Chem.* 276, 37861-37867.

10. Stroh A, Anderka O, Pfeiffer K, Yagi T, Finel M, Ludwig B and Schägger H (2004). Assembly of respiratory complexes I, III, and IV into NADH oxidase supercomplex stabilizes complex I in *Paracoccus denitrificans. J. Biol. Chem.* 279, 5000-5007.

11. Acŷin-Pérez R, Fernández-Silva M P, Moreno-Loshuertos P, Pérez-Martos R, Bruno A, Moraes C, and Enrŷiquez J A (2004). Respiratory complex III is required to maintain complex I in mammalian mitochondria. *Mol. Cell.* 13(6), 805–815.

Protocol No. 6.10: Determination of Gamma Radiation Effect on Oxidative Phosphorylation (OXPHOS) in Isolated Mitochondria of Irradiated Cells and its Modulation by Radioprotective Drug/ Compound: Measurement of Complex IV or Cytochrome C Oxidase Activity

Note

Cytochrome C oxidase is the terminal enzyme of the respiratory chain, that catalyzes electron transfer from cytochrome C to molecular oxygen. Cytochrome C oxidase is located in the mitochondrial inner membrane and is made up of 13 different subunits encoded by both mitochondrial and nuclear DNA. The catalytic core of the enzyme consists of three subunits all of which are mitochondrial gene products (Figure 18). Subunit II contains the binuclear CuA center that receives the electrons from cytochrome C. In subunit I, a low-spin heme (heme a) accepts electrons from the CuA and transfers them to a binuclear center consisting of a high-spin heme (heme a3) and a copper atom (CuB). Within the binuclear center, molecular oxygen is bound to heme a3 and sequentially reduced to water (1-8).

Assay Requirement

Cell culture facility, KH_2PO_4, sucrose, BSA, reduced cytochrome C, lauryl maltoside, KCN, spectrophotometer and gamma irradiator.

6.10.1 Assay Procedure

6.9.1.1 Collect exponentially growing cells (in medium changed a few hours before) by trypsinization and divided in following four groups:

Set A: Control cells (untreated)

Set B: Radioprotective drug treated cells

Set C: Cells irradiated by gamma radiation

Set D: Irradiated cells pretreated with radioprotective drug

Followed by drug and radiation treatment, cells were allowed to grow for 6-12 h with 5 per cent CO_2 concentration and 95 per cent humidity.

6.10.2 Measurement of Complex IV or Cytochrome C Oxidase Activity

6.10.2.1 Exponentially growing cells (in medium changed a few hours before) are collected by trypsinization, pelleted and resuspended in cold phosphate-buffered saline medium to a concentration of 5×10^6 cells/ml.

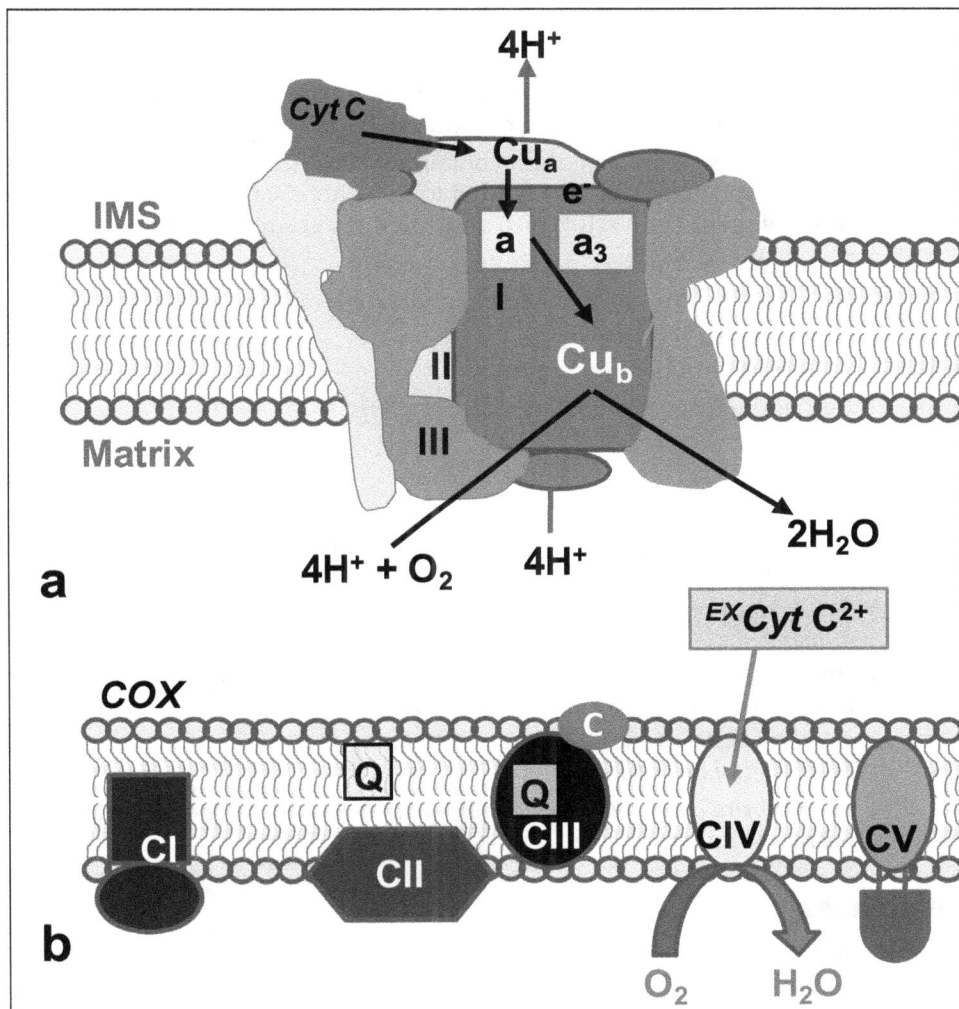

Figure 18: Complex IV or Cytochrome C Oxidase.
(a), Schematic representation of the complex showing, among others, the three subunits that form the catalytic core of the enzyme (subunits I, II, and III). The mechanism of electron transport and proton pumping is depicted; (b), Measurement of the cytochrome C oxidase activity (COX). ^{EX}Cyt C^{2+}, exogenous reduced cytochrome C.

6.10.2.2 The suspension was aliquoted and frozen at -80 °C until used for the different enzymatic assays.

6.10.2.3 The activity of the OXPHOS enzymes can be measured in isolated mitochondria.

6.10.2.4 A total of 20–40µg of mitochondrial protein will be sufficient to determine complex-IV or cytochrome C oxidase activity.

6.10.2.5 The assays described here are performed at 37°C (except the citrate synthase, at 30°C) in 1ml of medium.

6.10.2.6 The assay is performed at 550nm following the decrease in absorbance resulting from the oxidation of reduced cytochrome C.

6.10.2.7 In 1 ml of isosmotic medium containing 10mMKH$_2$PO$_4$ (pH 6.5), 0.25M sucrose and 1 mg/ml BSA, added to mitochondria (20–40µg of protein) or whole cells (3-10 x 10^4 cells) and 10µM reducedcytochrome C.

6.10.2.8 Follow the reaction for 3 minutes (rate 1).

6.10.2.9 Permeabilize the external mitochondrial membrane by adding 2.5mM lauryl maltoside (rate 2).

6.10.2.10 Follow the reaction for 3 minutes.

6.10.2.11 Inhibit the reaction with 240µMKCN (1-8).

Precuations

(i) *Dilute the samples as much as necessary until obtaining a good initial linear rate in the reaction for at least 2–3 minutes. Alternatively, the activity can be determined as a first-order rate constant.*

(ii) *Fresh cytochrome C solution must be reduced by adding some crystals of sodium dithionite.*

(iii) *When working with cells, it is recommended the preparation be freeze–thawed two or three times in isotonic medium before using. Also, it is suggested that the detergent be added after the reduced cytochrome C and the reaction started with the biological material.*

(iv) *The integrity of the outer mitochondrial membrane may be estimated by using the rate 1 and rate 2 values.*

References

1. Barrientos A (2002). *In vivo* and in organello assessment of OXPHOS activities. *Methods* 26 307–316.

2. Triepels R H, Van Den Heuvel L P, Trijbels J M and Smeitink J A (2001). Respiratory chain deficiency. *Am. J. Hum. Genet.* 106, 37–45.

3. Acin-Perez R, Fernandez-Silva P, Peleato M L, Perez-Martos A and Enriquez J A (2008). Respiratory active mitochondrial super-complexes. *Mol. Cell.* 32, 529–539.

4. Chretien D, Rustin P, Bourgeron T, Rotig A, Saudubray J M and Munnich A (1994). Reference charts for respiratory chain activities in human tissues. *Clin. Chim. Acta.* 228, 53–70.

5. Villani G, Greco M, Papa S and Attardi G (1998). Low reserve of cytochrome C oxidase capacity *in vivo* in the respiratory chain of a variety of human cell types. *J. Biol. Chem.* 273, 31829–31836.

6.   Schägger H and Pfeiffer K (2001). The ratio of oxidative phosphorylation complexes I–V in bovine heart mitochondria and the composition of Respiratory chain supercomplexes. *J. Biol. Chem.* 276, 37861-37867.

7.   Stroh A, Anderka O, Pfeiffer K, Yagi T, Finel M, Ludwig B and Schägger H (2004). Assembly of respiratory complexes I, III, and IV into NADH oxidase supercomplex stabilizes complex I in *Paracoccus denitrificans*. *J. Biol. Chem.* 279, 5000-5007.

8.   Li Y, Park J S, Deng J H, and Bai Y (2006). Cytochrome C oxidase subunit IV is essential for assembly and respiratory function of the enzyme complex. *J Bioenerg. Biom.* 38(5-6), 283–291.

Protocol No. 6.11: Determination of Gamma Radiation Effect on Oxidative Phosphorylation (OXPHOS) in Isolated Mitochondria of Irradiated Cells and its Modulation by Radioprotective Drugs: Measurement of Complex V or ATPase Activity

Introduction

Complex V or F1, F0-ATP synthase is the major enzyme responsible for the aerobic synthesis of ATP (Figure 19a). It comprises 16 different polypeptides, five of which constitute the membrane-extrinsic F_1-catalytic part (α_3; β_3; γ; d; e). The rest form the F_0 consisting of a set of hydrophobic proteins that are embedded in the membrane and that catalyze vectorial transfer of protons across the inner membrane, the direction being dependent on whether the enzyme is functioning in an ATP synthetic or hydrolytic mode. Four of the F0 subunits form a "stalk" that connects the F_1 and F_0 moieties. The ATP synthase acts as a rotary motor. The affinity change of substrates and products at catalytic sites is coupled to proton transport by rotation of the c subunit inside the α_3; β_3 oligomer.

Assay Requirement

Cell culture facility, lactate dehydrogenase, pyruvate kinase, NADH, defreezer, Tris, BSA, $MgCl_2$, KCl, carbonyl cyanide meta-chlorophenylhydrazone, antimycin A, Phosphoenol-pyruvate (PEP), ATP, spectrophotometer and gamma irradiator.

6.11.1 Assay Procedure

6.11.1.1 Harvested exponentially growing cells (in medium changed a few hours before) by trypsinization and divided in following four groups:

Set A: Control cells (untreated)

Set B: Radioprotective drug treated cells

Set C: Cells irradiated by gamma radiation

Set D: Irradiated cells pretreated with radioprotective drug

Followed by drug and radiation treatment, cells were allowed to grow for 6-12 h with 5 per cent CO_2 concentration and 95 per cent humidity.

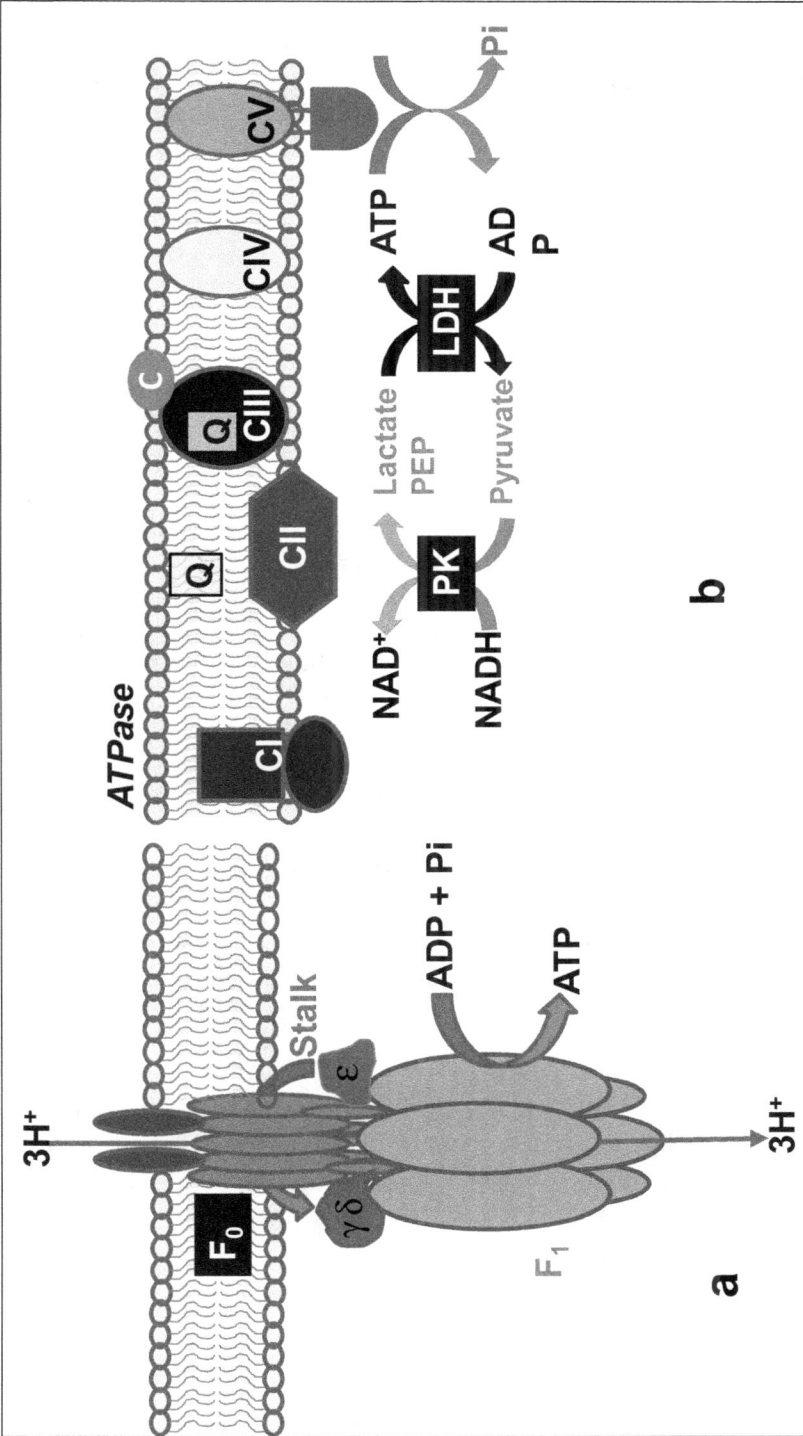

Figure 19: Complex V or F_1;F_0-ATP Synthase.

(a) Schematic representation of the complex; (b) Schematic representation of the complex. The ATP synthase acts as a rotary motor in the reaction; (b) Schematic representation of oligomycin-sensitive ATPase measurement by a coupled assay using lactate dehydrogenase (LDH) and pyruvate kinase (PK) as the coupling enzyme.

PEP: Phosphoenolpyruvate.

6.11.2 Measurement of Complex V (ATPase) Enzyme Activity

6.11.2.1 Exponentially growing cells (in medium changed a few hours before) was harvested by trypsinization, pelleted, and resuspended in cold phosphate-buffered saline medium to a concentration of 5×10^6 cells/ml.

6.11.2.2 The suspension was aliquoted and frozen at -80 °C until used for the different enzymatic assays.

6.11.2.3 The activity of the OXPHOS enzymes can be measured in isolated mitochondria.

6.11.2.4 A total of 20–40µg of mitochondrial protein will be sufficient to determine the activity of complex-V.

6.11.2.5 The assays described here are performed at 37°C (except the citrate synthase, at 30°C) in 1ml of medium.

6.11.2.6 Complex V activity can be measured spectrophotometrically by a coupled assay using lactate dehydrogenase and pyruvate kinase as the coupling enzyme.

6.11.2.7 The assay is performed at 340nm following the decrease in absorbance resulting from NADH reduction (Figure 19).

6.11.2.8 In a micro tube, prepared a medium containing 200ml 50mM Tris (pH 8.0), 5 mg/ml BSA, 20mM $MgCl_2$, 50mM KCl, 15mM carbonyl cyanide meta-chlorophenylhydrazone, 5mM antimycin A, 10mM phosphoenolpyruvate (PEP), 2.5mM ATP, 4 units of lactate dehydrogenase and pyruvate kinase, and 1mM NADH.

6.11.2.9 Incubate the medium for 5 minutes at 37 °C.

6.11.2.10 Incubate mitochondria (20–40µg of protein) in distilled water for 30s at 37 °C (use the spectrophotometer's chamber).

6.11.2.11 Add the medium to the chamber to start the reaction.

6.11.2.12 Follow the reaction for 3 minutes.

6.11.2.13 Add 3µM oligomycin and follow the reaction for an additional three minutes to distinguish the ATPase activity coupled to the respiratory chain (1-7).

Precautions

(i) A long incubation of the biological material in water will reduce the sensitivity to oligomycin by separation of the complex F0–F1.

(ii) It is not possible to follow this measurement in cells because of a strong oligomycin-insensitive ATPase activity.

(iii) PEP and NADH must be freshly prepared.

(iv) The expected inhibition with oligomycin is 60–90 per cent.

References

1. Barrientos A (2002). *In vivo* and in organello assessment of OXPHOS activities. *Methods* 26 307–316.

2. Acin-Perez R, Fernandez-Silva P, Peleato ML, Perez-Martos A and Enriquez JA (2008). Respiratory active mitochondrial super-complexes. *Mol. Cell.* 32, 529–539.

3. Fernandez-Vizarra E, Tiranti V and Zeviani M (2009). Assembly of the oxidative phosphorylation system in humans: what we have learned by studying its defects. *Biochim. Biophys. Acta.* 1793, 200–211.

4. Medja F, Allouche S, Frachon P, Jardel C, Malgat M, de Camaret B M, Slama A, Lunardi J, Mazat JP and Lombès A (2009). Development and implementation of standardized respiratory chain spectrophotometric assays for clinical diagnosis. *Mitochondrion.* 9(5), 331-339.

5. Cogliati S, Frezza C, Soriano M E, Varanita T, Quintana-Cabrera R, Corrado M, Cipolat S, Costa V, Casarin A, Gomes L C, Perales-Clemente E, Salviati L, Fernandez-Silva P, Enriquez J A and Scorrano L (2013). Mitochondrial cristae shape determines respiratory chain supercomplexes assembly and respiratory efficiency. *Cell.* 155 (1), 160–171.

6. Schägger H and Pfeiffer K (2001). The ratio of oxidative phosphorylation complexes I–V in bovine heart mitochondria and the composition of Respiratory chain supercomplexes. *J. Biol. Chem.* 276, 37861-37867.

7. Li Y, Park J S, Deng J H and Bai Y (2006). Cytochrome C oxidase subunit IV is essential for assembly and respiratory function of the enzyme complex. *J Bioenerg. Biom.* 38(5-6), 283–291.

Protocol No. 6.12: Determination of Gamma Radiation Induced Defects in Oxidative Phosphorylation in Muscles Tissue of Irradiated Animals and Role of Radioprotective Compounds to Ameliorate the Defects

Introduction

Defects in mitochondrial oxidative phosphorylation (OXPHOS) are frequent causes of severe inherited metabolic disorders and also contribute to aging. Similar types of disorder may induced by gamma radiation in irradiated cells/organisms though almost no information is exist on this topic due to no such study was conducted so far. Mitochondria severely affected organelle in the irradiated cell. Therefore, defect and perturbation in oxidative phosphorylation in irradiated cells cannot be denied. The OXPHOS system constitutes five multi-subunit complexes embedded in the mitochondrial inner membrane. Correct function of this system requires proper assembly of the ~80 proteins in the complexes, as well as numerous assembly factors. Blue native electrophoresis has become a crucial tool to investigate OXPHOS-related defects in mitochondrial disease patients. In addition, OXPHOS-assembly profiles can be obtained by two dimensional blue native/SDS gel electrophoresis, which provides additional information for identifying disease-causing mutations and insight in the role of specific proteins in the biogenesis of the OXPHOS system. Thus a practical guide on how to set-up the basic technique to study OXPHOS defects in irradiated cells/organism/patient-derived cells and tissues was discussed (1-6).

Assay Requirement

Cell culture facility, trypsin, culture medium, antibiotics, centrifuge, Teflon-glass Potter-Elvehjem homogenizer, MOPS sucrose buffer, ACBT-buffer, n-dodecyl β-D-maltoside, blue native sample buffer, coomassie brilliant blue G-250, electrophoresis system with power pack, peristaltic pump, gradient mixer, magnetic stirrer needle (18–20 Gauge), 40 per cent acryl amide/bis solution (29:1), blue native gel buffer 3X (composition: 1.5 M amino-caproic acid; 150 mM Bis–Tris; pH 7.0), glycerol, ammonium persulfate, TEMED, protein markers, anode buffer: (**composition:** 50 mM Bis–Tris; pH 7.0), cathode buffer A (composition: 50 mM Tricine, 15 mM Bis–Tris, 0.02 per cent serva blue G; pH 7.0), cathode buffer B (**composition:** 50 mM tricine, 15 mM Bis–Tris; pH 7.0), Tris-HCl, Tris, Glycine, $MgSO_4$, $Pb(NO_3)_2$, NADH, NTB Nitro Tetrazolium Blue (NBT), phenazine, methasulfate succinic acid, sodium phosphate buffer (PBS; pH 7.2), diamidobenzidine tetra hydrochloride (DAB), cytochrome C, ATP and gamma irradiator.

6.12.1 Assay Procedure

6.12.1.1 Muscles Tissue Used as Experimental Model

Experimental animals were divided into following four treatment groups:

Set A: Control animals (untreated)

Set B: Radioprotective drug treated animals

Set C: Animals irradiated by gamma radiation

Set D: Irradiated animals pretreated with radioprotective drug

Followed by drug and radiation treatment, muscle tissue of the experimental animals were collected at different time intervals.

6.12.1.1.1 Mince 20 mg of muscle tissue using a tissue chopper and further homogenize the tissue in ice-cold 250 µl MOPS sucrose buffer (composition: 440 mM sucrose, 20 mM Mops, 1 mM EDTA) using a tight fitted Teflon-glass Potter-Elvehjem homogenizer.

6.12.1.1.2 To obtain an enriched mitochondrial pellet, homogenate was centrifuge at 20,000xg for 20 minutes at 4 °C. This enriched mitochondrial pellet can be stored at -80 °C for several months.

6.12.1.1.3 Resuspend the mitochondria-containing pellet in 40µl ACBT buffer (composition: 75 mM Bis–Tris, 1.5 M aminocaproic acid, adjusted to pH 7.0 with HCl at 4°C) by swirling it with a small spatula.

6.12.1.1.4 Add 20µl n-dodecyl β-D-maltoside (10 per cent w/v in water) and leave the suspension for 10 minutes on ice.

6.12.1.1.5 Centrifuge the mixture at 20,000xg for 20 minutes at 4 °C. Transfer the supernatant in a fresh tube and measure protein concentration.

6.12.1.1.6 Add 10µl blue-native sample buffer (composition: 750 mM aminocaproic acid, 50 mM, Bis–Tris/HCl, pH 7.0, 0.5 mM EDTA) and mix gently. The sample was ready to load on the gel.

6.12.1.1.7 Load 10–40µg of protein on the gel.

6.12.1.2 When Skin Fibroblast Cells Used as Experimental Model

6.12.1.2.1 Harvest a minimum of one million cells using trypsin (approx. 175 cm2 flasks).

6.12.1.2.2 Bring the cell-suspension to a 15 ml tube and centrifuge the cells at 287g at 4 °C for 5 minutes.

6.12.1.2.3 Remove the supernatant, wash the pellet twice with cold PBS and centrifuge at 287g at 4 °C for 5 minutes.

6.12.1.2.4 Remove the supernatant and resuspend the cell-pellet in a concentration of 10^6 cells/100 µl cold PBS.

6.12.1.2.5 Transfer them into a 1.5 ml eppendorf tube and add 100 µl digitonin (4 mg/ml in PBS) and leave the cell-suspension on ice for 10 minutes (*Note: the optimal digitonin/protein ratio should be 0.8–1.6*).

6.12.1.2.6 Add 1 ml cold PBS (to dilute the digitonin).

6.12.1.2.7 Spin 10 minutes at 4 °C, at 10,000g/14000 rpm.

6.12.1.2.8 Remove the supernatant and wash the pellet twice with 1 ml cold PBS. Spin down again for 5 minutes at 4 °C using 10,000g/14000 rpm.

6.12.1.2.9 Remove the supernatant and continue with the crude mitochondrial pellet or snap freeze it in liquid nitrogen and store in a -80 °C freezer (These pellets can be kept for several months without any adverse effects).

6.12.1.2.10 The crude mitochondrial fraction is resuspended in 100 µl ACBT.

6.12.1.2.11 Add 20 µl 10 per cent β-dodecyl maltoside and homogenize the pellet by pipetting it up and down.

6.12.1.2.12 Leave the suspension on ice for 10 minutes.

6.12.1.2.13 Centrifuge the cells at 10,000 g at 4 °C for 30 minutes.

6.12.1.2.14 Put the supernatant in a new 1.5 ml eppendorf tube.

6.12.1.2.15 Measure the protein amount first (micro BCA protein assay kit, Pierce).

6.12.1.2.16 Add 10 µl of BN-Sample buffer.

6.12.1.2.17 Load 20–80 µg protein on the gel.

6.12.2 Blue Native Polyacrylamide Gel Electrophoresis

The casting of a gradient gel is the most difficult step of the present assay. For good separation of the OXPHOS protein complexes, 5–15 per cent gradient acrylamide gel should be used. But other gradients to zoom in on a specific OXPHOS complex can also be used, for instance a 10–18 per cent gradient can be used to better resolve complex IV subcomplexes.

NOTE: Ready-to-use blue native gels are commercially available can be used.

6.12.2.1 Casting of a Blue Native Gradient Gel

6.12.2.1.1 Assemble the gel sandwich; place it on the manual casting stand check for leakage using water.

6.12.2.1.2 Prepared the gel mixtures without TEMED and 10 per cent APS according to Table 2 and stored them on ice.

6.12.2.1.3 Rinsed gradient mixer and filled the tube with water.

6.12.2.1.4 Assemble gradient casting set-up. To reduce the dead volume in the system, use small diameter *e.g.* 0.8 or 1.6mm tubing.

6.12.2.1.5 Insert needle of gradient mixer between the glass plates. Place the needle to the side of the gel sandwich. Leave a few millimeters space between the end of the needle and the bottom of the sandwich to avoid unnecessary turbulence. This can affect the quality of the gradient. A piece of adhesive tape can be used to fix the needle at the top of the glass plates.

6.12.2.1.6 Add 10 per cent APS and TEMED according Table 2 to the gel-mix and immediately proceed with the casting of the gel.

6.12.2.1.7 Close the connecting valve between the compartments A and B of the gradient mixer and remove the water.

6.12.2.1.8 Fill compartment B with 3.5 ml of 5 per cent gel-mix and compartment A with 3.5 ml of the 15 per cent gel-mix. Store the remainder of the gel-mixes on ice; these can be used for a second gel.

6.12.2.1.9 Start the stirrer (do not set it at high speed) and open the valve between the compartments. Start the peristaltic pump at a speed of approximately 0.7 ml/minute. This way it will take about 10–12 minutes to cast the gel.

Note: faster speed of the pump cause turbulence, slower speed increases the risk of premature polymerization of the gels. Continue the casting of the gel until the tube (almost) does not contain gel-mix anymore; however, be careful not to let air-bubbles enter the gel.

6.12.2.1.10 Followed by filling the gel completly, stop the pump and gently remove the needle without disturbing the gradient.

6.12.2.1.11 Rinse the tubing immediately with water and repeat theprocedure for the second gel.

6.12.2.1.12 Let the gels polymerize at room temperature. Note: the water that was in the tubing will give a 2-4 mm layer ontop of the gel allowing a sharp boundary at the top of the gel.

6.12.2.1.13 The polymerization of the gels takes approximately 1-2h. *NOTE: leave the remainder of the gel-mixes on room temperature. This can be used to check whether the polymerization is completed or not.*

6.12.2.1.14 Remove the water from the top of the gel and proceed with the stacking gel.

Note: the gels can be stored for a few weeks. To prevent them from drying out, add 1X gel buffer on top of the gel and store the gel wrapped in cling film at 4°C.

6.12.2.1.15 Add 55µl 10 per cent APS and 5.5µl TEMED to the 4 per cent stacking gel mix and cast the stacking gel.

Table 2: Chemical Composition of Gradient Gel Components

Sl.No.	Components	Stacking Gel (4 per cent)	Gradient Gel (5 per cent)	Gradient Gel (15 per cent)
1.	Acrylamide solution (ml)	0.5 ml	1.25ml	3.75ml
2.	Gel buffer 3X (ml)	1.67ml	3.33 ml	3.33ml
3.	Glycerol (ml)	–	–	1.98ml
4.	H_2O (ml)	2.83 ml	5.42 ml	0.94ml
5.	APS(µl)	55 (µl)	60(µl)	30(µl)
6.	TEMED (µl)	5.5 (µl)	6.0 (µl)	3.0(µl)
7.	Total Volume (ml)	5 ml	10.0 ml	10.0 ml

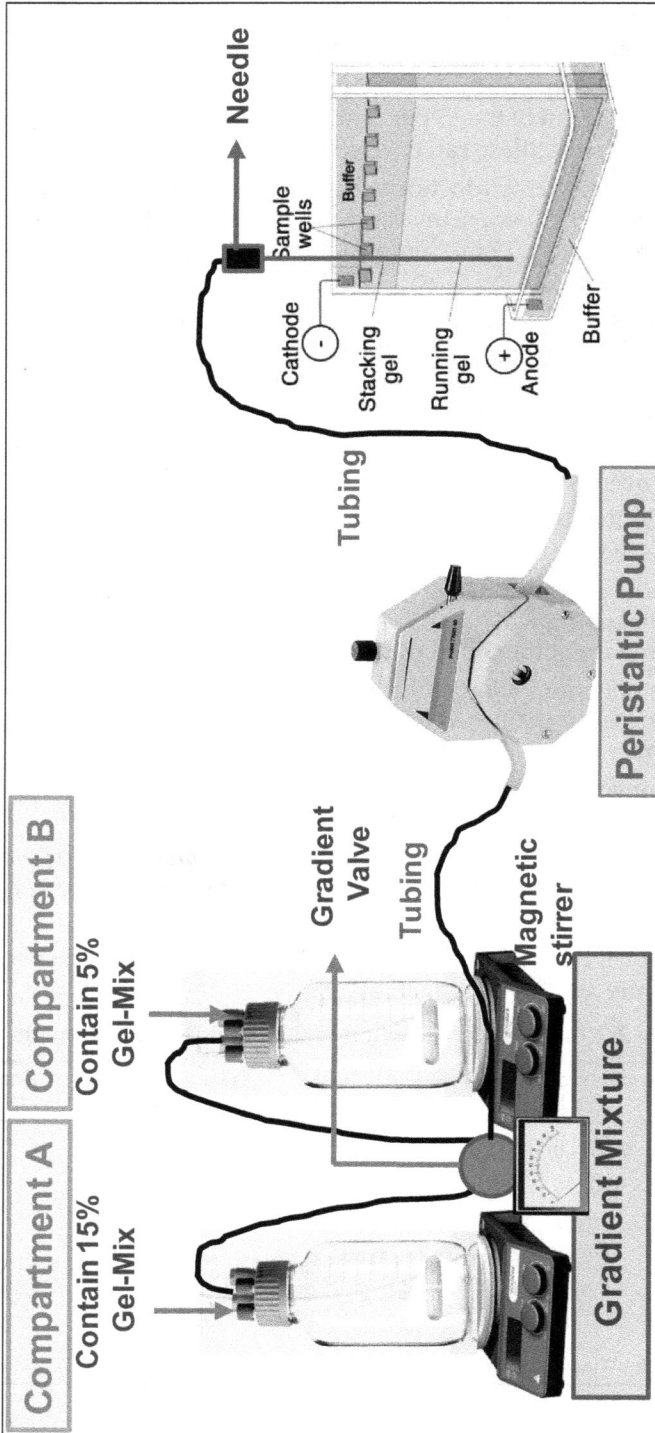

Figure 20: Schematic Representation of Gradient Gel Casting Process.

6.12.2.1.16 Place a comb (10 wells) between the glass plate and use 1.0 ml pipette filled-up the stacking around the comb (take care to avoid air-bubbles trapping under the comb).

6.12.3 Running of the Gel

6.12.3.1 Remove the comb from the stacking and rinse the wells with cathode buffer A (to remove un-polymerized acrylamide). Fillthe wells with cathode buffer A and gently load the samples (usually 20–40 µg of protein). Because of the higher density the sample will settle in the well (this way possible carry-over between the wells will be minimized).

6.12.3.2 Assemble the electrophoresis system and fill the inner compartment with cathode buffer A and the outer compartment with anode buffer.

6.12.3.4 Run electrophoresis for approximately 30 minutes at 30V then, raise the voltage up to 80V.

6.12.3.5 When front of the tracking dye has reached at the middle of the separating gel replace cathode buffer A by cathode buffer B.

6.12.3.6 Electrophoresis continues at 80V until the blue dye front reaches at the end of the gel.

6.12.3.7 Total running time will be approximately 4h.

6.11.4 Detection of OXHOS Complexes in Gels (*In situ* activity determination)

Gels are incubated at room temperature with following solutions:

6.12.4.1 **Complex I:** 5 mM Tris-HCl (pH 7.4), 0.1 mg/ml NADH and 2.5 mg/ml NTB (Nitro Tetrazolium Blue, Sigma).

6.12.4.2 **Complex II:** 5 mM Tris-HCl (pH 7.4), 0.2 mM phenazine metasulfate, 20 mM succinic acid and 2.5 mg/ml NTB.

6.12.4.3 **Complex IV**: 50 mM sodium phosphate buffer (pH 7.2), 0.5 mg/ml 3,3 0-diamidobenzidine tetra hydrochloride (DAB, Sigma-Aldrich) 0.05 mM cytochrome C (horse heart <95 per cent purity, Sigma).

6.12.4.4 **Complex V:** 35 mM Tris-270 mM Glycine (pH 8.3), 14 mM $MgSO_4$, 0.2 per cent $Pb(NO_3)_2$ and 8 mM ATP.

NOTE: Good signals are usually obtained after 2h for complex I and II and after overnight staining for the other complexes. After completion of the staining reaction, the gels were washed in water and immediately photographed or scanned. Incubation period with the specific colour reagent can be extended if desired.

6.12.5 Western Blotting of BN-PAGE

6.12.5.1 Western blotting of blue native gels can essentially be performed according standard procedures. It should be noted that since the native gradient gels are thicker than the standard SDS gels (1.5 mm instead of 0.75) blotting times should be extended. It takes at least 2-3h (or overnight in a cooled device) to assure complete transfer of the hydrophobic complexes. For antibody detection of the complexes it should be noted that some subunits give a poor detection, possibly due to the lack of epitope availability in the native complexes. Standard commercially available antibodies which yield quantitative results on blue native blots for the different complexes are:

Complex I. a-NDUFA9, a-NDUFS3

Complex II. a-70 kDa Fp

Complex III. a-core 2, a-FeS subunit

Complex IV. a-COX-I, a-COX-II, a-COX-IV

Complex V. a-ATPase subunit.

Standard western blotting procedure should be used for detection of the complex.

6.12.5.2 Good results were obtained using peroxidase-conjugated anti-mouse immunoglobulins as secondary antibodies and ECL1 plus reagents for signal detection (1-6).

References

1. Calvaruso M A, Smeitink J and Nijtmans L (2008). Electrophoresis techniques to investigate defects in oxidative phosphorylation. *Methods*. 46, 281–287.

2. Ding D, Enriquez-Algeciras M, Dave K R, Perez-Pinzon M and Bhattacharya S K (2012). Therole of deimination in ATP5b mRNA transport in a transgenic mouse model of multiple sclerosis. *EMBO Rep*. 13(3), 230-236.

3. Maalcke W J, de Almeida N M, Cirpus I, Gloerich J, Geerts W, Harhangi H R, Janssen-Megens E M, Francoijs K J, Stunnenberg H G, Keltjens J T, Jetten M S M and Strous M http://academic.research.microsoft.com/io.ashx?type=5&id=56631833&selfId1=0&selfId2=0&maxNumber=12&query= (2011). Molecular mechanism of anaerobic ammonium oxidation. *Nature*. 479(7371), 127-130.

4. Smits P, Saada A, Wortmann S B, Heister A J, Brink M, Pfundt R, Miller C, Haas D, Hantschmann R, Rodenburg R J T, Smeitink J A M and van den Heuvelhttp://academic.research.microsoft.com/io.ashx?type=5&id=56548331&selfId1=0&selfId2=0&maxNumber=12&query= L P (2011). Mutation in mitochondrial ribosomal protein MRPS22 leads to Cornelia de Lange-like phenotype, brain abnormalities and hypertrophic cardiomyopathy. *Eur J Human Genet*. 19(4), 394-399.

5. Iommarini L, Calvaruso M A, Kurelac I, Gasparre G and Porcelli A M (2013). Complex I impairment in mitochondrial diseases and cancer: Parallel roads leading to different outcomes. *Int. J. Biochem. Cell Biol*. 45(1), 47-63.

6. Hoefs S J G, Van Spronsen F J, Lenssen E W H, Nijtmans L G, Rodenburg R J, Smeitink J A M, Van-den Heuvel L P and Van-den Heuvel L P (2011). NDUFA10 mutations cause complex I deficiency in a patient with Leigh disease. *Europ. J. Hum. Genet*. 19(3), 270-274

Protocol No. 6.13: Assessment of Mitochondrial Outer Membrane Permeabilization (MOMP) during Apoptosis Induced by Gamma Radiation and its Protection using Radioprotective Agent in Irradiated Cells

Introduction

Mitochondria play a pivotal role in the regulation of apoptosis. An imbalance in apoptosis can lead to higher cell death. Unscheduled apoptosis has been linked to neuro-degeneration and other disorders, while inhibition of apoptosis can cause cancer.

Gamma radiation induced membrane lipid peroxidation, DNA damage, and protein/ enzyme oxidation and denaturation lead to apoptosis. An early and key event during apoptosis is the release of factors from mitochondria. In apoptosis, the mitochondrial outer membrane becomes permeable, leading to release of apoptogenic factors into the cytosol. One such factor, cytochrome C, is an electron carrier of the respiratory chain normally trapped within the mitochondrial inter-membrane space. Many apoptotic studies investigated mitochondrial outer membrane permeabilization (MOMP) by monitoring the release of cytochrome C. Here, we describe three reliable techniques that detect cytochrome C release from mitochondria, through subcellular fractionation or immunocytochemistry and fluorescence microscopy, or isolated mitochondria and recombinant Bax and t-Bid proteins in vitro (1-4).

Assay Requirement

Cell culture facility. Culture medium, trypsin, glass Dounce homogenizer (2ml) and tight B-type pestle, mini-electrophoresis and transfer unit, 1x phosphate buffered saline (PBS), mitochondrial isolation buffer (250 mM sucrose, 20 mM HEPES-KOH (pH 7.4), 10 mM KCl, 1.5 mM EGTA, 1.5 mM EDTA, 1 mMMgCl$_2$, 1 mM DTT), complete protease inhibitor cocktail, mitochondrial lysis buffer [(50 mM HEPES (pH 7.4), 1 per cent NP-40, 10 per cent glycerol, 1 mM EDTA, 2 mM DTT, protease inhibitor cocktail)], Quick start Bradford assay reagent, trans blot transfer medium, pure nitrocellulose (0.2 μm) membrane, Ponceau S solution, 12 per cent NuPAGE mini-gels plus running system, mouse anti-cytochrome C antibodies, rabbit anti-OMI/HtrA2 antibodies, west Pico Enhanced chemiluminescent substrate, rabbit anti-COX IV antibodies, mouse anti-β-Actin antibodies, restore Western blot stripping buffer, staurosporine, 2 mM stock in DMSO and gamma irradiator.

6.13.1 Assay Procedure

6.13.1.1 Subcellular Fractionation

6.13.1.1.1 Cells were cultured in appropriate medium and divided into following five experimental groups:

☆ Control cells (untreated)

☆ Cells irradiated with gamma radiation (2-4 Gy)

☆ Cells treated with radioprotective drug

☆ Irradiated cells pretreated with radioprotective drug

☆ Cells treated with apoptosis inducing drug staurosporine (1µM) in DMSO (Positive control) for 4h.

(NOTE: For apoptosis studies, it is important to harvest the non-adherent floating cells. Therefore, keep the cells directly on ice after staurosporine (STS) treatment without removing the medium, and remove the cells from the culture dish using a disposable cell scraper).

6.13.1.1.2 Cells under treatment were incubated for desired time intervals. Cells suspension was transfer into a 15 ml tube on ice. Cells were collected by centrifugation at 200xg for 5 minutes using a table top centrifuge at 4°C.

6.13.1.1.3 Cells pellets were washed twice with 10 ml of ice-cold PBS (pH 7.4). PBS must be removed from the cells pellet before adding the mitochondrial isolation buffer.

6.13.1.1.4 Pelletes were re-suspended with 300µl of ice-cold mitochondrial isolation buffer and incubated on ice for 20- 30 minutes.

6.13.1.1.5 Cells suspension was then transfered to a glass dounce with a tight pestle (B-type). 50–60 strokes were applied by moving the pestle gently up and down.

6.13.1.1.6 Cells homogenates were then transferred with a Pasteur pipette to an eppendorf tube and spined the sample at 800xg for 10 minutes at 4 °C to remove nuclei and unbroken cells.

6.13.1.1.7 Supernatant contains mitochondria was carefully removed, and transfer to a new eppendorf tube and spined at 22,000xg for 15 minutes at 4 °C.

6.13.1.1.8 The pellets contain mitochondria, were then resuspended with 100µl of mitochondrial lysis buffer.

6.13.1.1.9 Supernatant was save as the cytosolic fraction. You may save both mitochondrial and cytosolic fractions by freezing at -70 °C.

6.13.1.1.10 Protein concentration of the cytoplasmic and mitochondrial fraction was measured by using the Quickstart Bradford assay reagent and by comparing to a standard of BSA 0.05–0.5 mg/ml diluted in the mitochondrial isolation buffer.

6.13.1.2 Immunoblotting with Apoptogenic Factor Antibodies

6.13.1.2.1 Separate cytosolic and mitochondria-enriched fractions (20 μg protein) was analysed by SDS-PAGE and transfer to nitrocellulose membranes using standard procedure.

6.13.1.2.2 Followed by transfer, stain the nitrocellulose membrane with Ponceau S solution to confirm equal protein transfer.

6.13.1.2.3 Membranes was incubated for 2h at room temperature with PBS, 5 per cent nonfat milk, 0.1 per cent Tween 20 to block for nonspecific protein binding.

6.13.1.2.4 Primary mouse anti-cytochrome C(1:1000) or anti-OMI/HtrA2 (1:1000) antibodies were diluted in PBS, 0.1 per cent Tween 20, 5 per cent BSA and rock the blots overnight at 4 °C on a shaker.

6.13.1.2.5 The primary antibody solution was stored at 4 °C by adding 0.02 per cent NaN_3 and used repeatedly up to five times.

6.13.1.2.6 Next day, membranes were washed three times with PBS, 0.1 per cent Tween 20, for 5 minutes and incubated with secondary horseradish peroxidase-linked goat anti-mouse or anti-rabbit secondary antibodies (1:2000) for 2h at room temperature.

6.13.1.2.7 To confirm the equal protein loading, membranes was placed in the stripping buffer and rock for 5 minutes on a shaker.

6.13.1.2.8 After stripping, membrane was reblocked with PBS, 5 per cent nonfat milk, 0.1 per cent Tween 20 for 2h.

6.13.1.2.9 To confirm equal loading of cytosolic and mitochondrial fractions, mouse anti-β-Actin (1:10,000) and rabbit anti-COX IV (1:1000) antibodies were used.

6.13.1.2.10 Protein-antibody complexes were visualized, using enhanced chemiluminescence reagent (1-4).

Precautions

1. *To study MOMP as evidenced by cytochrome C release, we must have adequate amounts of healthy cells to start out.*

2. *Moreover, proper isolation of mitochondria using subcellular fractionation method may require a modest amount of digitonin (50 μg/ml).*

3. *Furthermore, optimization of the number of strokes you apply is critical to breaking the cells without disrupting the integrity of mitochondria.*

4. *Additionally, it is important to carefully remove supernatants without disturbing the mitochondrial pellet to avoid cross-contamination of the subcellular fractions.*

5. *Furthermore, different cell types may require an optimal time for staurosporine (STS) treatment/gamma radiation treatment. Therefore, one can treat cells for variable times (e.g., 3, 6, 9, 12 h) to stimulate apoptosis.*

References

1. Dave Z, Byfield M and Bossy-Wetzel E (2008). Assessing mitochondrial outer membrane permeabilization during apoptosis. *Methods* 46, 319–323.

2. http://jglobal.jst.go.jp/public/20090422/200902249073163838

3. Vempati U D, Torraco A and Moraes CT (2008). Mouse models of oxidative phosphorylation dysfunction and disease. *Methods*. 46(4), 241–247.

4. Calvaruso M A, Smeitink J and Nijtmans L (2008). Electrophoresis techniques to investigate defects in oxidative phosphorylation. *Methods*. 46, 281–287.

Protocol No. 6.14: Assessment of Mitochondrial Outer Membrane Permeabilization (MOMP) during Apoptosis Induced by Gamma Radiation and its Protection using Radioprotective Agent in Isolated Mitochondria

Assay Requirement

Cell culture facility, cell culture medium, trypsin, MSHE buffer:(260 mM mannitol, 70 Mm sucrose, 5 mM Hepes-KOH, pH 7.4), 1mM EGTA MSHA buffer: [(260 mM mannitol, 70 mM sucrose, and 5 mM HEPES-KOH (pH 7.6)], KCl buffer: (125 mM KCl, 4 mM MgCl$_2$, 5 mM Na$_2$HPO$_4$, 0.5 mM EGTA, 15 mM HEPES-KOH, pH 7.4), 5 mM succinate, 5 µM rotenone, recombinant Bax and t-BID protein, 4–12 per cent NuPAGE mini-gels, mouse anti-cytochrome C antibodies, rabbit anti Omi/ HtrA2, antibodies rabbit anti-VDAC antibodies, Rabbit anti-COX IV antibodies, quick start Bradford assay reagent, Western blot stripping buffer, bovine serum albumin (BSA), glass dounce homogenizer (40 ml) with teflon pestle (Bellco Glass), mini electrophoresis and transfer unit.

6.14 Assay Procedure

6.14.1 Cells were Cultured in Appropriate Medium and Divided into following Experimental Groups

- ☆ Control cells (untreated)
- ☆ Cells irradiated with gamma radiation (2-4 Gy)
- ☆ Cells treated with radioprotective drug
- ☆ Irradiated cells pretreated with radioprotective drug
- ☆ Cells treated with apoptosis induced drug staurosporine (STS) (1µM) in DMSO (Positive control) for 4h.

6.14.2 Preparation of Mouse Liver Mitochondria

6.14.2.1 After sacrifice the animals, the liver of a 4-to 6-week-old male mouse was rapidly excised and placed immediately into a 100 mm Petri dish on ice.

6.14.2.2 Special care should be taken not to disrupt and include the gall bladder in the preparation.

6.14.2.3 Liver tissue was minceed into small pieces using fine scissors.

6.14.2.4 Then 10 ml of ice-cold MSHE buffer containing BSA (1 mg/ml) was added.

6.14.2.5 Tissue was transfer with buffer mixed to the pre-chilled glass homogenizer with a Teflon pestle.

6.14.2.6 Apply sufficient numbers of strokes to homogenize the tissue completely, transfer the liver homogenate to centrifuge tubes and spun the samples at 3000xg for 2 minutes at 4 °C.

6.14.2.7 Remove the supernatant containing mitochondria and spin at 17500g for 3 minutes.

6.14.2.8 Wash the resulting mitochondrial pellet with MSHE without BSA and centrifuged at 17500xg for 5 minutes.

6.14.2.9 Finally, resuspended the pellet in 200 µl MSHA and stored on ice.

6.14.2.10 To measure mitochondrial protein concentration uses the Quick start Bradford assay reagent.

6.14.2.11 Remove 5µl from the final mitochondrial preparation and transfer to an eppendorf tube.

6.14.2.12 Spin the sample at maximum speed in an eppendorf centrifuge, discard the supernatant and resuspend the pellet in 20–50 µl mitochondrial lysis buffer and measure the protein concentration.

6.14.2.13 At this point, mitochondria are ready for use in the cytochrome C release assay.

6.14.3 *In vitro* Assay for Cytochrome C Release

6.14.3.1 Isolated mitochondria (100 µg) was incubated with recombinant 10 nM truncated Bid (t-Bid) and increasing oncentrations of Bax (5, 25, 50,100,150 nM) in 50 µlKCl buffer for 30 minutes at 30 °C.

6.14.3.2 After incubation, mitochondrial mixture was centrifuged at 13,000xg for 10 minutes at 4 °C.

6.14.3.3 To avoid contamination, 25 mm capillary loading tips was used to remove 45µl of the supernatant.

6.14.3.4 Discarded the remaining 5µl and then resuspended the mitochondrial pellets in 50µl mitochondrial lysis buffer.

6.14.3.5 Supernatant and mitochondrial fraction was separated by SDS–PAGE and transfer proteins on to nitrocellulose membranes by immunoblotting.

6.14.3.6 Protein transfer was verified by staining the membranes with Ponceau S solution.

6.14.3.7 Probed the transfer membranes with anti-cytochrome C antibodies, as described above.

6.14.3.8 To confirm the identity of the mitochondrial fraction, strip the membranes and reprobe the overnight at 4°C with Rabbit anti-COXIV or anti-VDAC antibodies (1:1000) as described above.

6.14.3.9 In a second experiment, release of cytochrome C or Omi/HtrA2 from mitochondria in the presence or absence of recombinant t-Bid (10 nM) and a constant amount of Bax (150 nM) at different time points including 5, 15, and 30 minutes was studied at 30°C.

6.14.3.10 The maximal cytochrome C release may be observed only at 30 minutes (1-4).

References

1. Dave Z, Byfield M and Bossy-Wetzel E (2008). Assessing mitochondrial outer membrane permeabilization during apoptosis. *Methods* 46, 319–323.

2. http://jglobal.jst.go.jp/public/20090422/200902249073163838

3. Vempati U D, Torraco A and Moraes C T (2008). Mouse models of oxidative phosphorylation dysfunction and disease. *Methods*. 46(4), 241–247.

4. Calvaruso M A, Smeitink J and Nijtmans L (2008). Electrophoresis techniques to investigate defects in oxidative phosphorylation. *Methods*. 46, 281–287.

Protocol No. 6.15: DCDF Assay for Measurement of Mitochondrial Reactive Oxygen Species in Isolated Mitochondria of Irradiated Cells/Tissues and their Inhibition by Radioprotective Drug Treatment

Introduction

Mitochondria are generators of reactive oxygen species (ROS) in the cells and tissues. ROS include, predominantly, the superoxide anion (O^{--2}), hydrogen peroxide (H_2O_2), and the hydroxyl radical (OH^*). H_2O_2 is the most stable and abundant, because it is a by-product of superoxide dismutase (SOD) catalyzed reaction. Nitroxy radicals also contribute to mitochondrial ROS. Present method focused on the investigation of mitochondrial production of H_2O_2 and discusses new developments in the measurement of lipid peroxidation. In general, measurements of mitochondrial ROS production can be carried out either in vitro, using conventional preparations of isolated mitochondria or in live cells. Unfortunately, several methods originally applied to detect and investigate mitochondrial ROS are not suitable for cellular studies. Measuring mitochondrial ROS is complicated not only due to methodological problems, but also by common views on the relative contribution of individual redox groups that may lead to incorrect interpretations of data (1-5).

Assau Requirement

Cell culture facilities, Opaque 96-well microtiter plate, phosphate-buffered saline (PBS) buffer, DCFDA, dimethyl sulfoxide (DMSO), succinate, horseradish peroxidase, automated shaker, hydrogen peroxide, fluorescence spectrophotometer, gamma irradiator.

6.15.1 Assay Procedure

6.15.1.1 Culture the cells in appropriate medium and divided into following experimental groups:

☆ Control cells (untreated)

☆ Cells irradiated with gamma radiation (2-4 Gy)

☆ Cells treated with radioprotective drug

☆ Irradiated cells pretreated with radioprotective drug

6.15.1.2 In a typical assay, each well of a 96-well microtiter plate (made of material optically suitable for fluorescence measurements) was filled with phosphate-buffered saline (PBS) containing 1μM of DCFDA diluted from a stock solution in dimethyl sulfoxide (DMSO) and 0.5 mg/ml of sub mitochondrial particles or mitochondria, to a final volume of 0.15–0.2 ml.

6.15.1.3 The reaction was started by the addition of mitochondrial substrates, *e.g.*, 10mM succinate and the mixture incubated with automated shaking for at least 30 minutes at 30 °C.

6.15.1.4 Measurements were routinely done using excitation wavelength 485nm and emission wavelength 520nm with 5-nm bandwiths.

6.15.1.5 Samples can be further supplemented with exogenous peroxidases, *i.e.*1 unit/ml horseradish peroxidase (HRP) to enhance DCFDA oxidation several fold.

6.15.1.6 Inhibitors, when used, are generally added in the reaction mixture well before addition of the biological preparation (mitochondrial preparation), to minimize DCFDA oxidation by the radicals produced with endogenous substrates.

6.15.1.7 For quantitative evaluation of the DCF fluorescence generated from the reaction mixture should by plotted on standard curve prepared using commercial DCF.

6.15.1.8 These calibrations can remain linear in the ranges spanning sub-nanomolar to sub-micromolar concentrations of DCFDA, depending on the electronic and optical capacity of the instrument available.

6.15.1.9 Since HRP and other peroxidases also increase the spontaneous oxidation of DCFDA without mitochondria, and the method is extremely sensitive, measurements can be achieved in absence of external peroxidases.

6.15.1.10 In general, DCFDA measurements are more than 80 per cent sensitive to hydrogen peroxide and partially sensitive to the hydroxyl radical.

6.15.1.11 The contribution of other ROS, including organic hydroperoxides to DCFDA oxidation in the presence of mitochondria is generally minimal.

6.15.2 ROS Measurements in Live Cells

6.15.2.1 Cells were cultured in appropriate medium and divided into following experimental groups:

☆ Control cells (untreated)

☆ Cells irradiated with gamma radiation (2-4 Gy)

☆ Cells treated with radioprotective drug

☆ Irradiated cells pretreated with radioprotective drug

6.15.2.2 96-well plate was supplemented with cells 0.5–5 x10^5/ml in PBS (or another appropriate medium), and incubated with 1–2 μM DCDFA (diluted in the medium from a 5mM stock solution in DMSO) for 10-15 minutes in the dark.

6.15.2.3 Measurements were usually initiated using excitation at 485–495nm and emission at 520–530 nm by the addition of nutrients like glutamine and glucose and can follow appropriate manipulations of mitochondrial or cell function. Fluorescence intensity was recorded at every 1 or 2 minutes for at least 30 minutes.

6.15.2.4 Blanks containing equivalent concentrations of dead cells (prepared, most conveniently, by freezing and thawing an aliquot of the cell suspension were required to evaluate the background oxidation of DCFDA.

6.15.3 ROS-specific Staining of Mitochondria

6.15.3.1 To achieve effective ROS-sensitive staining, cells (0.4-1×10^6/ml) were suspended at in fresh growth medium containing 0.5 µM CM-H_2X Ros Molecular probes. H_2XRos probe should be prepared just before the experiment by dissolving the content of a commercial vial in 0.1ml of pure DMSO.

6.15.3.2 Cells were then incubated in H2X Ros staining medium for 15 minutes at room temperature, washed twice with PBS, and then fixed with conventional procedures (*e.g.*, with a fresh solution of 3.7 per cent formaldehyde in PBS.

6.15.3.3 After being washed and mounted on slides, the stained cells are analyzed by cyto-fluorescence techniques.

6.15.3.4 Confocal microscopy should be undertaken with intermediate photomultiplier voltage and a 590-nm band pass filter for fluorescence emission.

6.15.3.5 The basal fluorescence intensity may vary significantly in different types of cells and under different conditions of stimulating ROS production (1-5).

References

1. Esposti M D (2002). Measuring mitochondrial reactive oxygen species. *Methods*. 26, 335–340.

2. Esposti M D and McLennan H (1998). Mitochondria and cells produce reactive oxygen species in virtual anaerobiosis: relevance to ceramide-induced apoptosis. *Methods*. 430 (3), 338–342.http://dx.doi.org/10.1016/S0014-5793(98)00688-7

3. Wanga H and Josepha J A (1999). Quantifying cellular oxidative stress by dichlorofluorescein assay using microplate reader. *Free Rad. Biol. Med.* 27, 612–616.

4. Reinier M J, Van Golen R F and Van- Gulik T M (2012). 22,73-Dichlorofluorescein is not a probe for the detection of reactive oxygen and nitrogen species. *J. Hepatol.* 56 (5), 1214–1216.

5. Myhre O, Andersen J M, Aarnes H and Fonnum F (2003). Evaluation of the probes 22,72-dichlorofluorescin diacetate, luminol, and lucigenin as indicators of reactive species formation. *Biochem. Pharmacol.* 65 (10), 1575–1582.

Protocol No. 6.16: Flow Cytometry Based Assays for the Measurement of Apoptosis-Associated Mitochondrial Membrane Depolarisation and Cytochrome C Release in Irradiated Cells and its Protection by Radioprotective Agent

Introduction

Mitochondria play a pivotal role in life and death of the cells because they produce majority of energy required for survival and also regulate the intrinsic pathway of apoptosis. The involvement of mitochondria in cell death is generally measured by mitochondrial membrane depolarisation or mitochondrial outer membrane permeabilisation (MOMP). These events can be assayed using cationic dyes that are attracted to the negative charge across the inner membrane of healthy mitochondria or by following translocation of cytochrome C from the mitochondria to the cytoplasm respectively. These events progress rapidly in individual cells but are observed as bi-phasic peaks in flow cytometry assays because cell death generally occurs asynchronously in a population. This allows researchers to use flow cytometry to easily distinguish healthy cells with intact mitochondria from dying cells with permeabilised mitochondria.

Assay Requirement

Cell cultutre facility, digitonin, KCl, PBS (pH7.4), trypan blue, paraformaldehyde, saponin, BSA, purified anti-cytochrome C, fluorescent conjugate corresponding secondary antibody, anti-mouse-IgG1-PE, purified mouse anti-cytochrome C, TMRE, carbonyl cyanide 4 (trifluoromethoxy) phenylhydrazone (FCCP), Carbonyl cyanide m-chlorophenylhydrazone (CCCP), 2,4-dinitrophenol (DNP), centrifuge with 96 well plate holder, florescence spectrophotometer, gamma irradiator.

6.16,1 Assay Procedure

6.16.1.1 Measuring Cytochrome C Release by Flow Cytometry

6.16.1.1.1 Cells were cultured in appropriate medium and divided into following experimental groups:

- ☆   Control cells (untreated)
- ☆   Cells irradiated with gamma radiation (2-4 Gy)
- ☆   Cells treated with radioprotective drug
- ☆   Irradiated cells pretreated with radioprotective drug

6.16.1.2 Step 1. Permeabilising the Plasma Membrane

6.16.1.2.1 2×10^4–10^5 cells in 100 µl ice-cold permeabilisation buffer (100 mMKCl, 50–200 µg/ml digitonin in PBS) were re-suspended and incubated on ice for 3–5 minutes until 95 per cent of the cells permeabilised (*e.g.* stain positive with trypan blue).

6.16.1.3 Step 2. Fixing the Cells

6.16.1.3.1 100µl paraformaldehyde (4 per cent in PBS) aws directly added to the permeabilised cells (50:50 v/v) and immediately pelleted in 96 well v-bottom plate by centrifugation for 5 minutes at 500xg at 4 °C.

6.16.1.3.2 Supernatant was gently removed and incubated the cells in paraformaldehyde (4 per cent) for 20 minutes at room temperature.

6.16.1.3.3 Cells were washed three times in 200 µl PBS (700xg for 5 minutes at 4 °C).

6.16.1.3.4 Paraformaldehyde (PFA) was directly added to the cells in permeabilization buffer contained diluted digitonin. Cell fixation process was initiated.

6.16.1.3.5 Subsequent addition of PFA ensures complete fixation of the cells.

6.16.1.4 Step 3. Immunocytochemistry and Flow Cytometry

6.16.1.4.1 Resuspend the cells in 200 µl blocking buffer (0.05 per cent Saponin, 3 per cent BSA in PBS) and incubate for 1h at room temperature.

6.16.1.4.2 Add anti-cytochrome C (1:200) antibodies and incubate overnight at 4 °C.

6.16.1.4.3 Wash the cells three times in 200 µl PBS (700xg for 5 minutes) and add fluorescent conjugate secondary antibody diluted in blocking buffer (1:200) and incubated for 1 h at room temperature.

6.16.1.4.4 Wash the cells three times in 200 µl PBS (700xg for 5 minutes) and analyse the samples by flow cytometry using appropriate excitation and emission filters for the conjugated fluorochrome.

6.16.1.5 Steps 4. Measuring Mitochondrial Membrane Depolarization by Flow Cytometry

6.16.1.5.1 Add TMRE directly to the cells (5×10^5/ml) in media. After that a protonophore such as carbonyl cyanide 4 (trifluoromethoxy) phenylhydrazone (FCCP, 5 µM final) was added to an untreated sample as a positive control.

6.16.1.5.2 Cells were incubated for 5 minutes at 37 °C and measure the level of fluorescence in the cells by flow cytometry using appropriate excitation and emission filters.

6.16.1.5.3 TMRE is optimally excited at 549nm and emits at 574 nm but can also be excited using a 488 nm laser available on most flow cytometers.

6.16.1.5.4 Cells with healthy mitochondria will have high fluorescence and cells with depolarized mitochondria have low fluorescence.

6.16.1.5.5 Temperature can directly impact TMRE staining, therefore it is important that do not place the cells on ice prior to analysing them on the flow cytometer.

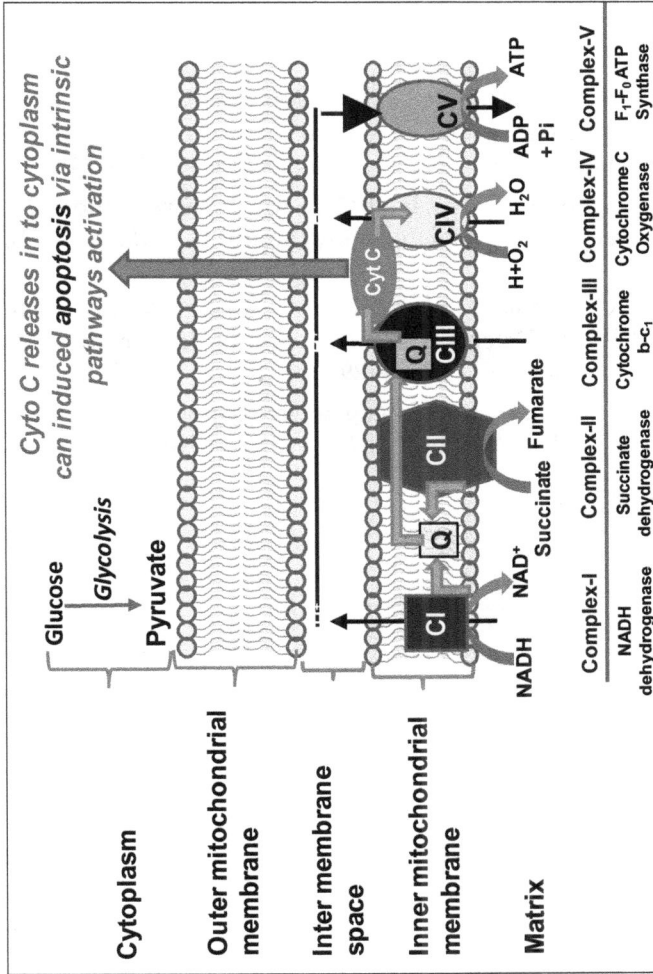

Figure 21: Mitochondrial Membrane Potential Alteration Mechanism.

The mitochondrial membrane potential is generated by protons that are transferred from the matrix to the inter-membrane space (thick black arrows) as electrons are passed down electron transport chain (thick grey arrows). Glucose is converted to pyruvate by glycolysis, which is taken up by mitochondria and used as fuel for the Kreb's cycle. A critical step in this process involves conversion of succinate to fumarate by succinate dehydrogenase (Complex II). This liberates an electron that carried to complex III by co-enzyme Q. Electrons are also generated by NADH dehydrogenase (Complex I) and passed to coenzyme Q. These electrons are carried by cytochrome C to cytochrome C oxygenase (Complex IV), which converts hydrogen and oxygen to water, a final step in the electron transport chain. Protons are passed across the inner membrane during the reactions in complex I, III and IV, creating a negative charge across the mitochondrial inner membrane. This charge is then used by ATP synthase to fuse ADP and inorganic phosphate to ATP.

6.16.1.5.6 FCCP will depolarise mitochondria of healthy cells indicating the level of fluorescence that can be expected in cells that have lost of mitochondrial potential during apoptosis.

6.16.1.5.7 If you do not observe a difference between the control and FCCP treated sample, the assay should be refined using the information. Carbonyl cyanide m-chlorophenylhydrazone (CCCP) or 2,4-dinitrophenol (DNP), can also be used in place of FCCP (1-4).

References

1. Christensen M E, Jansen E S, Sanchez W and Waterhouse N J (2013). Flow cytometry based assays for the measurement of apoptosis associated mitochondrial membrane depolarisation and cytochrome c release. *Methods* 61, 138–145.

2. Renaulta T T, Florosa K V and Chipuk J E (2013). BAK/BAX activation and cytochrome c release assays using isolated mitochondria. *Methods*. 61(2), 146–155.

3. Berghe T V, Grootjans S, Goossens V, Dondelinger Y, Krysko D V, Takahashi N and Vandenabeele P (2013). Determination of apoptotic and necrotic cell death *in vitro* and *in vivo*. *Methods*. 61(2), 117–129.

4. Martin S J and Henry C M (2013). Distinguishing between apoptosis, necrosis, necroptosis and other cell death modalities. *Methods*. 61(2), 87–89.

Protocol No. 6.17: Evaluation of ATP Synthesis in Mitochondria Isolated from Animal Tissues/Cells Irradiated by Gamma Radiation and its Modulation by Radioprotective Drug Pretreatment

Assay Requirement

Cell culture facilities, buffer H, (D-mannitol, sucrose, HEPES, EGTA, and bovine serum albumin (BSA), pH 7.2, glass-Teflon pestle, centrifuge, electrode oxygraph, succinate or glutamate, malate, ADP, luciferase/luciferin system and gamma irradiator.

6.17.1 Assay Procedure

6.17.1.1 Animals or Cells under Evaluation were Divided into following Four Groups

Gp.1 Control animals/cells (untreated cells),

Gp.2 Irradiated animals/cells (cells treated with gamma radiation),

Gp.3 Animals/Cells treated with radioprotective drug,

Gp.4 Irradiated Animal/Cells pretreated with radioprotective drug,

Note: Alternatively isolated mitochondria may also be used to make experimental groups as mentioned above.

6.17.1.2 Isolation of Mitochondria

6.17.1.2.1 A whole mouse liver/cells was homogenized with 8ml of buffer H, containing 0.22 M D-mannitol, 0.07M sucrose, 20mM HEPES, 1mM EGTA, and 1 per cent bovine serum albumin (BSA) (pH 7.2), buffer R (contained sucrose, HEPES, $MgCl_2$, EGTA, KH_2PO_4 (pH 7.4) with a glass-Teflon pestle gently by hand with approximately 5 strokes.

6.17.1.2.2 The homogenate was centrifuged at 1500xg for 5 minutes.

6.17.1.2.3 The supernatant was centrifuged at 10,000xg for 10 minutes.

6.17.1.2.4 The mitochondrial pellet was resuspended in a small volume (approximately 250 µl) of buffer H and kept on ice.

6.17.1.2.5 The final protein concentration was 40–60 mg/ml. Lower dilutions are not recommended because they tend to cause uncoupling of mitochondrial respiration.

6.17.1.2.6 Mitochondria from a mouse brain were extracted by homogenization in 2ml of buffer H.

6.17.1.2.7 The homogenate was centrifuged at 1500xg for 5 minutes.

6.17.1.2.8 The supernatant was kept on ice, while, the resulting pellet resuspended in 1 ml of buffer H.

6.17.1.2.9 Supernatant was subjected to a second centrifugation at 1500g.

6.17.1.2.10 Supernatants were combined and centrifuged at 13,500xg for 10 minutes.

6.17.1.2.11 Mitochondrial pellet was resuspended in 50–100 µl of buffer H and kept on ice.

6.17.1.2.12 The final protein concentration was set at 20–40 mg/ml.

6.17.1.2.13 All centrifugation steps were performed at 4 °C.

6.17.1.3 Measurement of ATP Synthesis

6.17.1.3.1 Prior to measuring ATP synthesis, the coupling state of mitochondria was tested by polarography.

6.17.1.3.2 Approximately 600 µg of liver and 400 µg of brain mitochondrial proteins were resuspended in 0.3 ml of buffer R (containing 0.25M sucrose, 50mM HEPES, 2mM $MgCl_2$, 1mM EGTA, 10mM KH_2PO_4, pH 7.4).

6.17.1.3.3 Oxygen consumption rate was measured in a Clark-type electrode oxygraph using either 20 mM succinate or 30 mM glutamate plus 30mM malate in the absence of exogenous ADP (state 2 respiration) and after addition of 300mM ADP (state 3 respiration).

6.17.1.3.4 The addition of 1 mg/ml oligomycin inhibits the mitochondrial ATP synthase and reduces oxygen consumption to a rate equal to that prior to the addition of ADP.

6.17.1.3.5 Respiratory control ratios (RCRs, the ratio between state 2 and state 3 respirations) should be above 3 and 7 in liver and 2.5 and 4 in brain with succinate and glutamate/malate, respectively.

6.17.1.3.6 Lower RCRs indicate uncoupling of mitochondria.

6.17.1.3.7 30µg of liver and 100 µg of brain mitochondrial proteins were resuspended in buffer A and ATP synthesis was measured with the luciferase/luciferin system (1-8).

References

1. Fern-andez-Vizarra E, Lopez-Perez M J, and Enriquez J A (2002). Isolation of biogenetically competent mitochondria from mammalian tissues and cultured cells. *Methods* 26 292–297.

2. Hornig-DoH T, Günther G, Bust M, Lehnartz P, Bosio A and Wiesner R J (2009). Isolation of functional pure mitochondria by superparamagnetic microbeads. *Anal Biochem*. 389(1), 1-5.

3. Wallace D C (1999). Mitochondrial diseases in man and mouse. *Science*. 283(5407), 1482-1488.

4. Sergeant N, Wattez A, Galvn-valencia M, GhestemA, P David J, Lemoine J Sautire E P, Dachary J, Mazat J P, Michalski J C, Velours J and Mena-lpez R (2003). Association of ATP synthase α-chain with neurofibrillary degeneration in alzheimer's disease. *Neuroscience*. 117(2), 293-303.

5. Trifunovic A (2006). Mitochondrial DNA and ageing. *Biochimica Et Biophysica Acta-bioenergetics*. 1757(5), 611-617.

6. Frezza C, Cipolat S and Scorrano L (2007). Organelle isolation: functional mitochondria from mouse liver, muscle and cultured filroblasts. *Nature Protocols*. 2(2), 287-295.

7. Kiss K, Brozik A, Kucsma N, Toth A, Gera M, Berry L, Vallentin A, Vial H, Vidal M and Szakacs G (2012). Shifting the paradigm: The putative mitochondrial protein ABCB6 resides in the lysosomes of cells and in the plasma membrane of erythrocytes. *PLOS One*. 7(5), e3737.

8. José Miguel P. de Oliveira F and de Graaff L H (2011). Proteomics of industrial fungi: Trends and insights for biotechnology. *Appl. Microbiol. Biotechnol*. 89 (2), 225-237.

Section 7

Immune System Radioprotective Efficacy Evaluation of Radioprotective Compounds

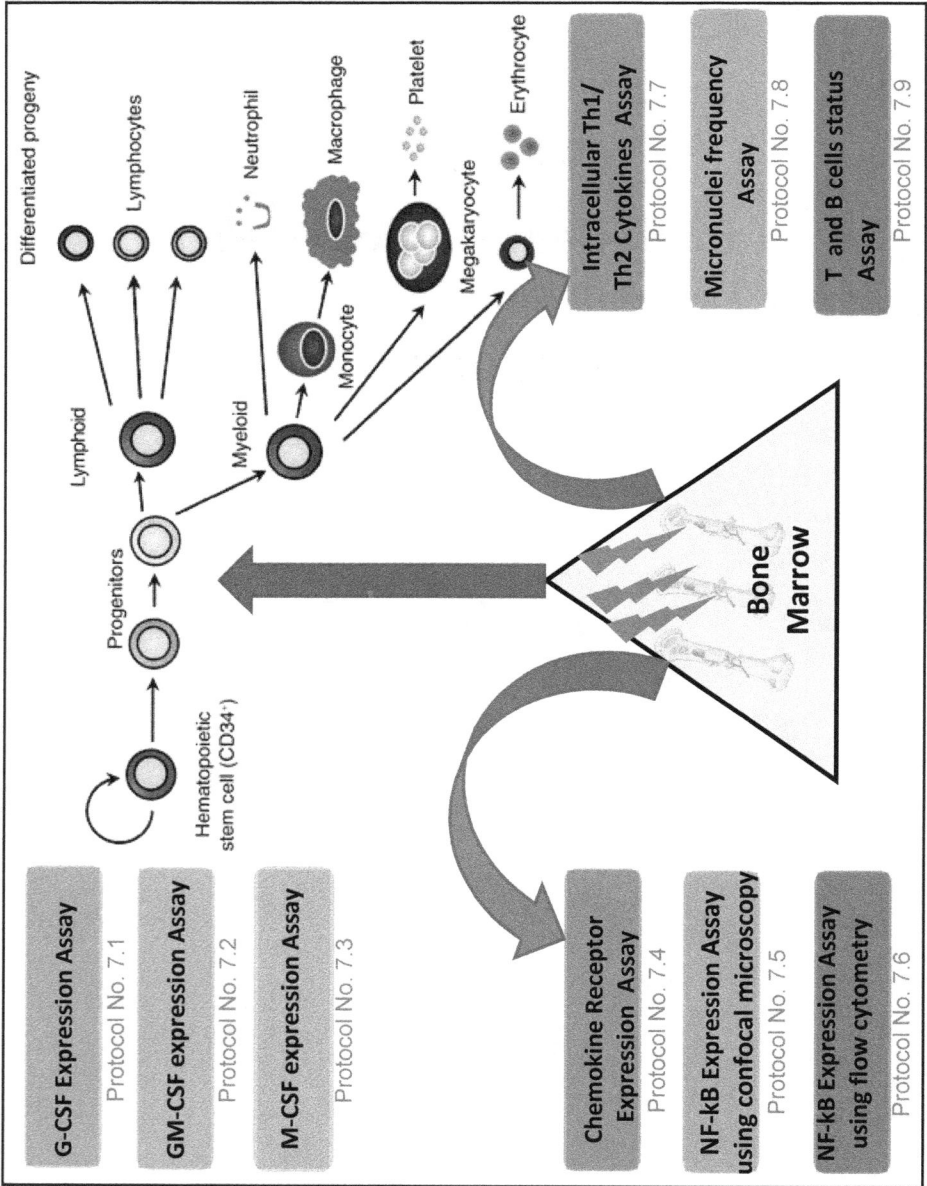

Figure 22: Schematic Respresentation of Integration of Different Protocols Associated with Hematopoitic System Radioprotection.

Protocol No. 7.1: Determination of G-CSF Expression in Irradiated Cells/Animals and Modulation by Radioprotective Compounds Preatreatment using ELISA Assay

Assay Requirement

Cell culture facility, culture medium, trypsin, culture plates/flask, 96 microtiter ELISA plates, ammonium chloride, potassium bicarbonate, EDTA, sodium chloride, NP-40, PBS (pH 7.2), NH_4Cl, $KHCO_3$, Triton X-100, Tris HCl, EGTA, cell strainer, 26 gauge needle, G-CSF Elisa kits, refrigerated centrifuge and gamma irradiator.

7.1.1 Assy Procedure

7.1.1.1 Six animals kept in each group and following treatment groups were formulated or alternatively animals/human cells can be used:

Group 1: Control Animals: (n=6); animals administered with PBS (0.5 ml; i.p.)

Group 2: Cells/Animals: (n=6); irradiated with gamma radiation (2-10 Gy)

Group 3: Cells/Animals (n=6) treated with radioprotective drug

Group4: Cells/Animals (n=6) treated with radioprotective drug before irradiation (2-10 Gy)

7.1.1.2 Followed by final treatments, mice were kept back at experimental animal facility. The animals euthanized at different time intervals and bone marrow, spleen and blood samples were collected and processed to obtain cell free supernatant.

7.1.1.3 Serum Sample Preparation

7.1.1.3.1 Blood samples were collected from the mice treated with gamma radiation and radioprotective drug at different time intervals using retero-orbital blood collection method.

7.1.1.3.2 Blood was left for 30 minutes at room temperature and then centrifuged at 2500xg.

7.1.1.3.3 Serum was collected and kept at -20°C till used.

7.1.1.4 Spleen Single Cell Preparation

7.1.1.4.1 Spleen and bone marrow samples were collected from experimental animals at different time intervals (6-72h).

7.1.1.4.2 Spleen was crushed through frosted slide and single cell suspension prepared using cell strainer.

7.1.1.4.3 Single cell suspension was centrifuged to 4000Xg.

7.1.1.4.4 Cell pellet obtained was washed with sterile PBS (pH 7.4).

7.1.1.4.5 Cell suspension in PBS was stored at 4°C temperature.

7.1.1.5 Bone Marrow Single Cell Preparation

7.1.1.5.1 To prepare bone marrow single cell suspension, femurs were excised from both ends.

7.1.1.5.2 Muscles were removed through surgical procedure.

7.1.1.5.3 Bone marrow was flushed with sterile PBS through 26 guaze needle.

7.1.1.5.4 Single cell suspension was collected into sterile tube (1.5 ml) and stored at 4°C temperature.

7.1.1.6 Preparation of Cell Lysate

7.1.1.6.1 Single cell suspension prepared from spleen or bone marrow was centrifuged at 4000Xg at 4°C.

7.1.1.6.2 Cells pallet obtained were treated with RBC lysis buffer (composition: NH_4Cl 0.82 per cent, $KHCO_3$ 0.1 per cent and 1.8mg EDTA 3.6 per cent in water) and incubated for 10 minutes with mild shaking at 37°C temperature.

7.1.1.6.3 The mixture was then centrifuged at 4000Xg. Cell pellets obtained were rewashed with 1.0 ml PBS.

7.1.1.6.4 The cell suspension again centrifuged at 4000Xg and cell pellets was treated with cell lysis buffer (1 per cent Triton X-100, 10mM Tris HCl, 150mM NaCl, 0.5 per cent NP-40, 1mM EDTA and 0.2 per cent EGTA) supplemented with protease inhibitors.

7.1.1.6.5 Cells were then incubated in ice for 30 minutes and centrifuged at 10000Xg for 30 minutes at 4°C temperature.

7.1.1.6.6 Supernatant was separated and processed for analysis of G-CSF.

7.1.1.7 Estimation of G-CSF Expression using Elisa Method

7.1.1.7.1 G-CSF level in serum, spleen and bone marrow was analyzed using Elisa kit.

7.1.1.7.2 Standard and samples (equivalent to 100µg protein) were added to the 96 well plates pre-coated with the anti-mouse G-CSF monoclonal antibodies and kept at room temperature for 2.5h with gentle shaking.

7.1.1.7.3 Plates then washed with the wash buffer and 100µl of biotinylated secondary antibodies were added to each well. The plates were then left at room temperature for 1hr.

7.1.1.7.4 After washing, 100µl of streptavidin was added to each well of the plate containing samples. Plates were then left in incubation at 37° for 45 minutes.

7.1.1.7.5 Plates were washed with washing buffer and 100µl of TMB added to each well and plate placed in the dark for half an hour.

7.1.1.7.6 After 30 minutes of incubation, stop solution was added to terminate the reaction.

7.1.1.7.7 Absorbance was read at 415nm.

7.1.1.7.8 Quantitative estimation of G-CSF expression was carried out by calculating the corresponding quantity on the standard curve prepared using G-CSF standard supplied along with the kit (1-9).

References

1. Layton J E, Shimamoto G, Osslund T, Hammacher A, Smithi D K, Treutlein H R, and Boone T (1999). Interaction of granulocyte colony-stimulating factor (G-CSF) with its receptor. *J. Biol. Chem.* 274 (25), 17445–17451.

2. Aoki Y, Sha S, Mukai H and Nishi Y (2000). Selective stimulation of G-CSF gene expression in macrophages by a stimulatory monoclonal antibody as detected by a luciferase reporter gene assay. *J. Leukocyte Biol.* 68(5), 757-764.

3. Kumar J, Fraser F W, Riley C, Ahmed N, McCullochD R and WardA C (2014). Granulocyte colony-stimulating factor receptor signalling via Janus kinase 2/ signal transducer and activator of transcription 3 in ovarian cancer. *Br. J.Cancer* 110, 133–145.

4. Bohlius J, Reiser M, Schwarzer G and Engert A (2003). Impact of granulocyte colony-stimulating factor (CSF) and granulocyte-macrophage CSF in patients with malignant lymphoma: a systematic review. *Br. J. Haematol.* 122, 413–423.

5. Brandstetter T, Ninci E, Bettendorf H, Perewusnyk G, Stolte J, Herchenbach D, Sellin D, Wagner E, Kochli O R and Bauknecht T (2001). Granulocyte colony-stimulating factor (G-CSF) receptor gene expression of ovarian carcinoma does not correlate with G-CSF caused cell proliferation. *Cancer* 91, 1372–1383.

6. Gutschalk C M, Herold-Mende C C, Fusenig N E and Mueller M M (2006). Granulocyte colony-stimulating factor and granulocyte-macrophage colony-stimulating factor promote malignant growth of cells from head and neck squamous cell carcinomas *in vivo. Cancer Res.* 66, 8026–8036.

7. Munstedt K, Hackethal A, Eskef K, Hrgovic I and Franke F E (2010). Prognostic relevance of granulocyte colony-stimulating factor in ovarian carcinomas. *Arch. Gynecol. Obstet.* 282, 301–305.

8. Ward A C (2007). The role of the granulocyte colony-stimulating factor receptor (G-CSF-R) in disease. *Front Biosci.* 12, 608–618.

9. Ward A C, Hermans M H A, Smith L, Van Aesch Y M, Schelen A M, Antonissen C, Touw I P (1999a). Tyrosine-dependent and independent mechanisms of STAT3 activation by the human granulocyte colony-stimulating factor (G-CSF) receptor are differentially utilized depending on G-CSF concentration. *Blood.* 93, 113–124.

Protocol No. 7.2: Determination of GM-CSF Expression in Irradiated Cells and its Modulation by Radioprotective Compounds Preatreatment using ELISA Assay

Assay Requirement

Cell culture facility, culture medium, trypsin, culture plates/flask, 96 microtiter ELISA plates, ammonium chloride, potassium bicarbonate, EDTA, EGTA, Triton X-100, Tris HCl, sodium chloride, NP-40, PBS (pH 7.2), NH_4Cl, $KHCO_3$, EDTA, Triton X-100, cell strainer, 26 gauge needle, GM-CSF Elisa kits, refrigerated centrifuge and gamma irradiator.

7.2.1 Assay Procedure

7.2.1.1 Six animals kept in each group and following treatment groups were formulated or alternatively animals/human cells can be used:

Group 1: Animals: (n=6); animals administered with PBS (0.5 ml; i.p.)

Group 2: Animals: (n=6); irradiated with gamma radiation (2-10 Gy)

Group 3: Animals (n=6) treated with radioprotective drug

Group 4: Animals (n=6) treated with radioprotective drug before irradiation (2-10 Gy)

7.2.1.2 Followed by final treatments, mice were kept back at experimental animal facility. The animals euthanized at different time intervals and bone marrow, spleen and blood samples were collected and processed to obtain cell free supernatant.

7.2.1.2 Serum Sample Preparation

7.2.1.2.1 Blood samples were collected from the mice treated with gamma radiation and radioprotective drug at different time intervals using retero-orbital blood collection method.

7.2.1.2.2 Blood was left for 30 minutes at room temperature and then centrifuged at 2500xg.

7.2.1.2.3 Serum was collected and kept at -20°C till used.

7.2.1.3 Spleen Single Cell Preparation

7.2.1.3.1 Spleen and bone marrow samples were collected from experimental animals at different time intervals (6-72h).

7.2.1.3.2 Spleen was crushed through frosted slide and single cell suspension prepared using cell strainer.

7.2.1.3.3 Single cell suspension was centrifuged to 4000Xg.

7.2.1.3.4 Cell pellet obtained was washed with sterile PBS (pH 7.4).

7.2.1.3.5 Cell suspension in PBS was stored at 4°C temperature.

7.2.1.4 Bone Marrow single cell preparation

7.2.1.4.1 To prepare bone marrow single cell suspension, femurs were excised from both ends.

7.2.1.4.1 Muscles were removed through surgical procedure.

7.2.1.4.1 Bone marrow was flushed with sterile PBS through 26 guaze needle.

7.2.1.4.1 Single cell suspension was collected into sterile tube (1.5 ml) and stored at 4°C temperature.

7.2.1.5 Preparation of Cell Lysate

7.2.1.5.1 Single cell suspension prepared from spleen or bone marrow was centrifuged at 4000Xg at 4°C.

7.2.1.5.2 Cells pellet obtained were treated with RBC lysis buffer (composition: NH_4Cl 0.82 per cent, $KHCO_3$ 0.1 per cent and 1.8mg EDTA 3.6 per cent in water) and incubated for 10 minutes with mild shaking at 37°C temperature.

7.2.1.5.3 The mixture was then centrifuged at 4000Xg. Cell pellets obtained were rewashed with 1.0 ml PBS.

7.2.1.5.4 The cells suspension again centrifuged at 4000Xg and cell pellets was treated with Cell lysis buffer (1 per cent Triton X-100, 10mM Tris HCl, 150mM NaCl, 0.5 per cent NP-40, 1mM EDTA and 0.2 per cent EGTA) supplemented with protease inhibitors.

7.2.1.5.5 Cells were then incubated in ice for 30 minutes and centrifuged at 10000Xg for 30 minutes at 4°C temperature.

7.2.1.5.6 Supernatant was separated and processed for analysis of GM-CSF.

7.2.1.6 Estimation of GM-CSF Expression using Elisa Method

7.2.1.6.1 GM-CSF level in serum, spleen and bone marrow was analyzed using Elisa kit.

7.2.1.6.2 Standard and samples (equivalent to 100μg protein) were added to the 96 well plates pre-coated with the anti-mouse GM-CSF monoclonal antibodies and kept at room temperature for 2.5h with gentle shaking.

7.2.1.6.3 Plates then washed with the wash buffer and 100µl of biotinylated secondary antibodies were added to each well. The plates were then left at room temperature for 1hr.

7.2.1.6.4 After washing, 100µl of streptavidin was added to each well of the plate containing samples. Plates were then left in incubation at 37°C for 45 minutes.

7.2.1.6.5 Plates were washed with washing buffer and 100µl of TMB added to each well and plate placed in the dark for half an hour.

7.2.1.6.6 After 30 minutes of incubation, stop solution was added to terminate the reaction.

7.2.1.6.7 Absorbance was read at 415nm.

7.2.1.6.8 Quantitative estimation of GM-CSF expression was carried out by calculating the corresponding quantity on the standard curve prepared using GM-CSF standard supplied along with the kit (1-11).

References

1. Bronte V, Chappell D B, Apolloni E, Cabrelle A, Wang M, Hwu P and Restifo N P (1999). Unopposed production of granulocyte-macrophage colony-stimulating factor by tumors inhibits CD8+ T cell responses by dysregulating antigen-presenting cell maturation. *J. Immunol.* 162, 5728–5737.

2. Ueda Y, Kondo M and Kelsoe G (2005). Inflammation and the reciprocal production of granulocytes and lymphocytes in bone marrow. *J. Exp. Med.* 201, 1771–1780.

3. Rutella S, Zavala F, Danese S, Kared H and Leone G (2005). Granulocyte colony-stimulating factor: a novel mediator of T cell tolerance. *J. Immunol.* 175, 7085–7091.

4. Bronte V (2009). Myeloid-derived suppressor cells in inflammation: Uncovering cell subsets with enhanced immunosuppressive functions. *Eur. J. Immunol.* 39, 2670–2672.

5. Disis M L, Bernhard H, Shiota F M, Hand S L, Gralow J R, Huseby E S, Gillis S and Cheever M A (1996). Granulocyte-macrophage colony-stimulating factor: an effective adjuvant for protein and peptide-based vaccines. *Blood.* 88, 202.

6. Jager E, Ringhoffer M, Dienes H P, Arand M, Karbach J, Jager D, Ilsemann C, Hagedorn M, Oesch F and Knuth A. (1996). Granulocyte-macrophage-colony-stimulating factor enhances immune responses to melanoma-associated peptides *in vivo*. *Int. J. Cancer* 67: 54.

7. Okada E, Sasaki S, Ishii N, Aoki I, Yasuda T, Nishioka K, Fukushima J, Wahren B, and Okuda K (1997). Intranasal immunization of a DNA vaccine with interleukin 12 and granulocyte macrophage colony stimulating factor (GM-CSF) expressing plasmids in liposomes induce strong mucosal and cell-mediated immune responses against HIV-1 antigen. *J. Immunol.* 159, 3638.

8.   Haworth C, Brennan F M, Chantry D, Turner M, Maini RN and Feldmann M (1991). Expression of granulocyte-macrophage colony-stimulating factor in rheumatoid arthritis: Regulation by tumor necrosis factor-α. *Eur. J. Immunol.* 21 (10), 2575–2579.

9.   Zsengellér Z K, Reed J A, Bachurski C J, LeVine A M, Forry-Schaudies S, Hirsch R and Whitsett J A. (1998). Adenovirus-mediated granulocyte-macrophage colony-stimulating factor improves lung pathology of pulmonary alveolar proteinosis in granulocyte-macrophage colony-stimulating factor-deficient mice. *Hum Gene Ther.* 9(14), 2101-2109.

10. Reed J A, Ikegami M, Robb L, Begley C G, Ross G and Whitsett J A (2000). Distinct changes in pulmonary surfactant homeostasis in common beta-chain- and GM-CSF-deficient mice. *Am J. Physiol. Lung Cell Mol. Physiol.* 278(6), 1164-1171.

11. Ikegami M, Jobe A H, Huffman Reed J A and Whitsett J A (1997). Surfactant metabolic consequences of overexpression of GM-CSF in the epithelium of GM-CSF-deficient mice. *Am. J. Physiol.* 273(4 Pt 1), 709-714.

Protocol No. 7.3: Determination of M-CSF Expression in Irradiated Cells and its Modulation by Radioprotective Compounds Preatreatment using ELISA Assay

Assay Requirement

Cell culture facility, culture medium, trypsin, culture plates/flask, 96 wells microtiter ELISA plates, ammonium chloride, potassium bicarbonate, EDTA, EGTA, Triton X-100, Tris HCl, sodium chloride, NP-40, PBS (pH 7.2), NH_4Cl, $KHCO_3$, cell strainer, 26 gauge needle, M-CSF Elisa kits, refrigerated centrifuge and gamma irradiator.

7.3.1 Assy Procedure

7.3.1.1 Six animals kept in each group and following treatment groups were formulated or alternatively animals/human cells can be used:

Group 1: Control Animals: (n=6); animals administered with PBS (0.5 ml; i.p.)

Group 2: Animals: (n=6); irradiated with gamma radiation (2-10 Gy)

Group 3: Animals (n=6) treated with radioprotective drug

Group 4: Animals (n=6) treated with radioprotective drug before irradiation (2-10 Gy)

7.3.1.2 Followed by final treatments, mice were kept back at experimental animal facility. The animals euthanized at different time intervals and bone marrow, spleen and blood samples were collected and processed to obtain cell free supernatant.

7.3.1.3 Serum Sample Preparation

7.3.1.3.1 Blood samples were collected from the mice treated with gamma radiation and radioprotective drug at different time intervals using retero-orbital blood collection method.

7.3.1.3.2 Blood was left for 30 minutes at room temperature and then centrifuged at 2500xg.

7.3.1.3.3 Serum was collected and kept at -20°C till used.

7.3.1.4 Spleen Single Cell Preparation

7.3.1.4.1 Spleen and bone marrow samples were collected from experimental animals at different time intervals (6-72h).

7.3.1.4.2 Spleen was crushed through frosted slide and single cell suspension prepared using cell strainer.

7.3.1.4.3 Single cell suspension was centrifuged to 4000xg.

7.3.1.4.4 Cell pellet obtained was washed with sterile PBS (pH 7.4).

7.3.1.4.5 Cell suspension in PBS was stored at 4°C temperature.

7.3.1.5 Bone Marrow Single Cell Preparation

7.3.1.5.1 To prepare bone marrow single cell suspension, femurs were excised from both ends.

7.3.1.5.2 Muscles were removed through surgical procedure.

7.3.1.5.3 Bone marrow was flushed with sterile PBS through 26 guaze needle.

7.3.1.5.4 Single cell suspension was collected into sterile tube (1.5 ml) and stored at 4°C temperature.

7.3.1.6 Preparation of Cell Lysate

7.3.1.6.1 Single cell suspension prepared from spleen or bone marrow was centrifuged at 4000Xg at 4°C.

7.3.1.6.2 Cells pallet obtained were treated with RBC lysis buffer (composition: NH_4Cl 0.82 per cent, $KHCO_3$ 0.1 per cent and 1.8mg EDTA 3.6 per cent in water) and incubated for 10 minutes with mild shaking at 37°C temperature.

7.3.1.6.3 The mixture was then centrifuged at 4000Xg. Cell pellets obtained were rewashed with 1.0 ml PBS.

7.3.1.6.4 The cell suspension again centrifuged at 4000Xg and cell pellets was treated with Cell lysis buffer (1 per cent Triton X-100, 10mM Tris HCl, 150mM NaCl, 0.5 per cent NP-40, 1mM EDTA and 0.2 per cent EGTA) supplemented with protease inhibitors.

7.3.1.6.5 Cells were then incubated in ice for 30 minutes and centrifuged at 10000Xg for 30 minutes at 4°C temperature.

7.3.1.6.6 Supernatant was separated and processed for analysis of M-CSF.

7.3.1.7 Estimation of M-CSF Expression using ELISA Method

7.3.1.7.1 M-CSF in the serum, spleen and bone marrow was analyzed using ELISA kit.

7.3.1.7.2 Standard and samples (100µl, 1:8 dilutions with assay buffer) were added to the plates pre-coated with the mouse M-CSF and left at room temperature for 2h with gentle shaking.

7.3.1.7.3 After decanting the solution from the plate, 100µl of biotinylated secondary anti-mouse M-CSF antibody was added to each well.

7.3.1.7.4 Plates were than left at room temperature for 1h. Followed by incubation completion, plates were washed three times with PBS (pH 7.4).

7.3.1.7.5 100µl of HRP-streptavidin solution was added to each well of the plate.

7.3.1.7.6 Plates were then incubated at 37°C temperature for 45 minutes and washed with washing buffer three times again.

7.3.1.7.7 100µl of TMB added to each well and plates were kept in the dark for 30 minutes.

7.3.1.7.8 Followed by incubation, reaction was stopped by adding stop solution (provided alongwith the kit).

7.3.1.7.9 Colour developed was read using UV-Vis spectrometer at 415nm.

7.3.1.7.10 Quantitative estimation of M-CSF expression was carried out by calculating the corresponding quantity on the standard curve prepared using M-CSF standard supplied along with the kit (1-11).

References

1. Murphy Jr. G M, Yang L and Cordell B (1998). Macrophage colony-stimulating factor augments β-amyloid-induced interleukin-1, interleukin-6, and nitric oxide production by microglial cells. *J. Biol. Chem.* 273, 20967-20971.

2. Walker F, Zhang H H, Matthews V, Weinstock J, Nice E C, Ernst M, Rose-John S and Burgess A W (2008). IL6/sIL6R complex contributes to emergency granulopoietic responses in G-CSF and GM-CSFdeficient mice. *Blood*, 111(8), 3978-3985.

3. Gruber M F, Weih K A, Boone E J, Smith P D and Clouse K A (1995). Endogenous macrophage CSF production is associated with viral replication in HIV-1-infected human monocyte-derived macrophages. *J. Immunol.* 154(10), 5528-5535.

4. Tanabea N, Maeno M, Suzukic N, Fujisakia K, Tanaka H, Ogisoa B and Itoe K (2005). IL-1α stimulates the formation of osteoclast-like cells by increasing M-CSF and PGE$_2$ production and decreasing OPG production by osteoblasts. *Life Sciences.* 77(6), 615–626.

5. Fujita K and Janz S (2007). Attenuation of WNT signaling by DKK-1 and -2 regulates BMP2-induced osteoblast differentiation and expression of OPG, RANKL and M-CSF. *Mol. Cancer* 6, 71-75.

6. Eubank T D, Galloway M, Montague C M, Waldman W J and Marsh C B (2003). M-CSF induces vascular endothelial growth factor production and angiogenic activity from human monocytes. *J. Immunol.* 171(5), 2637-2643.

7. Okazaki T, Ebihara S, Takahashi H, Asada M, Kanda A and Sasaki H (2005). Macrophage colony-stimulating factor induces vascular endothelial growth factor production in skeletal muscle and promotes tumor angiogenesis. *J. Immunol.* 174 (12). 7531-7538.

Protocol No. 7.4: Flow Cytometric Analysis of Chemokine Receptor Expression in Cerebro-Spinal Fluid and Peripheral Blood Leukocytes upon Gamma Irradiation and its Modulation by Radioprotective Drug Pretreatment

Note

Collection of cerebrospinal fluid (CSF) for diagnostics purposes from the lumbar sub-arachnoid space is a routine procedure in clinical neurology. The capability to study individual cells in samples with low cell numbers has made flow cytometry an important tool for chemokine receptor expression analysis. Several methodological variables such as cell staining, temperature and cell isolation etc. techniques may affect the final outcome of the procedure. The normal lumbar CSF is relatively a cellular, containing fewer than 5 leukocytes/μl, predominantly T lymphocytes and a few myeloid cells, whereas, dramatically increased numbers of leukocytes can be detected in the CSF during CNS infection. Analysis of CSF cells by flow cytometry provides an opportunity to characterize the surface phenotype of CSF cells directly ex vivo. By careful handling of the cells, we have been able to use flow cytometry to compare the expression of chemokine receptors on cells of the samples from peripheral blood and CSF in irradiated animals and can correlate the outcome with inflammatory response in irradiated and radioprotective drug treated animals (1-.8).

Assay Requirement

Heparin, FACS lysing solution, cold FACS buffer, paraformaldehyde (PFA), Monoclonal antibodies CCR1, CCR2, CCR3, CCR4, CCR5, CCR6, CCR7, CXCR3, IgG1, IgG2a, and IgG2b isotype controls, anti-IgG1 fluorescein isothiocyanate (FITC), anti-IgG2b PE and anti-IgG2a biotin, and streptavidin-PE-Cy7, 1.5ml polypropylene Eppendorf tubes, centrifuge, FACS analyzer, gamma irradiator.

7.4.1 Assay Procedure

7.4.1.1 Whole blood staining using directly conjugated mAbs

7.4.1.1.1 Animals were divided into four experimental groups:

    i. Control untreated mice

    ii. Mice treated with Radioprotective drug

    iii. Irradiated mice

    iv. Irradiated mice pretreated by radioprotective drug

Followed by final treatments, micro were kept at experimental animal facility. The animals were euthanized at different time interval and CSF/blood collected for further analysis.

7.4.1.1.2 100µl of heparinized venous blood was blocked with 0.2 mg/ml normal mouse IgG for 15 minutes at room temperature in 5.0ml polystyrene FACS tubes.

7.4.1.1.3 Directly, desired conjugated monoclonal antibodies (mAbs) (as mentioned in the assay requirement section) were added according to pre-determined optimal concentrations without washing out the blocking reagent. Samples were then incubated for 15 minutes at room temperature in the dark.

7.4.1.1.4 After staining, 2.0ml FACS lysing solution was added to lyse erythrocytes. Samples were kept for 10 minutes at room temperature in the dark. Samples were then centrifuged at 4°C for 7 minutes at 600xg. The supernatant removed and palletes were washed 2-3 times using cold FACS buffer (2.0 ml), followed by centrifugation at 4 °C for 7 minutes at 600xg.

7.4.1.1.5 Supernatant was aspirated and 100µl of 1 per cent paraformaldehyde (PFA) added and incubated for 5 minutes at room temperature. Finally, the samples were centrifuged at 4 °C for 7 minutes at 600xg. The supernatant removed and the cells were re-suspended in FACS buffer.

7.4.1.2 Whole Blood Staining using Unconjugated mAbs

7.4.1.2.1 100µl of heparinized venous blood was blocked with 0.2 mg/ml normal IgG from the same species as the secondary anti-IgG antibodies for 15 minutes at room temperature.

7.4.1.2.1 Primary unconjugated monoclonal abtibodies (mAbs) were added according to predetermined optimal concentrations without washing out the blocking reagent. Samples were incubated for 15 minutes at room temperature and washed with ice-cold FACS buffer.

7.4.1.2.1 Cells were resuspended in 100µl FACS buffer and secondary anti-IgG antibodies mixed to predetermined concentrations. Samples were incubated for 15 minutes at room temperature in the dark, followed by two washes with ice-cold FACS buffer.

7.4.1.2.1 Directly conjugated mAbs against cell lineage markers were added in a third staining step. After incubation for 15 minutes at room temperature in the dark, 2.0 ml FACS lysing solution was added to the samples. Washing and fixation were performed as described above.

7.4.1.3 Staining of CSF

7.4.1.3.1 10ml aliquots of cerebrospinal fluid (CSF) were collected using standered procedure, immediately placed on ice, and centrifuged at 600xg for 7 minutes at 4 °C. Excess supernatant *i.e.* 100µl was removed and cells were re-suspended in 1.5 ml polypropylene Eppendorf tubes.

7.4.1.3.2 Nonspecific FcR binding was inhibited by the incubation with 0.2 mg/ml normal mouse IgG (directly conjugated mAbs) or normal IgG from the same species for 15 minutes at room temperature.

7.4.1.3.3 Primary mAbs (mentioned in assay requirement section) were added in identical concentration as for whole blood stainings without washing out the blocking reagent. Samples were incubated for 15 minutes at room temperature in the dark.

7.4.1.3.4 Cells were washed once with 1.0 ml ice-cold FACS buffer and centrifuged at room temperature for 30s at 5000xg. The supernatant was carefully removed. Cells were resuspended in 100 µl of FACS buffer and secondary anti-IgG antibodies added for 15 minutes at room temperature in the dark. Samples were washed once as described above and directly conjugated mAbs against cell lineage markers added in a third step.

7.4.1.3.1 After the final wash and removal of the supernatant, 100µl of 1 per cent paraformaldehyde (PFA) was added for 5 minutes at room temperature in the dark. Finally, the samples were centrifuged for 30s at 5000xg, the supernatant was removed, and cells were resuspended in FACS buffer.

7.4.1.3.1 Flow cytometric analysis was carried out using standard procedure to identify the leucocyte specific markers in CSF or blood samples (1-15).

References

1. Kivisaakk P, Liu Z, Trebst C, Tucky B, Wu L, Stine J, Mack M, Rudick R A, Campbell J J and Ransohoff R M (2003). Flow cytometric analysis of chemokine receptorexpression on cerebrospinal fluid leukocytes. *Methods* 29, 319–325.

2. Berhanu D, Mortari F, De Rosa S C and Roederer M (2003). Optimized lymphocyte isolation methods for analysis of chemokine receptor expression. *J. Immunol. Methods.* 279(1-2), 199-207.

3. Anselmo A, Mazzon C, Borroni E M, Bonecchi R, Graham G J and Locati M (2014). Flow cytometry applications for the analysis of chemokine receptor expression and function. *Cytometry A.* 85(4), 292-301.

4. Nieto J C, Cantó E, Zamora C, Ortiz M A, Juárez C, *et al.* (2012) Selective loss of chemokine receptor expression on leukocytes after cell isolation. *PLoS ONE* 7(3): e31297. doi:10.1371/journal.pone.0031297.

5. Dux R, Kindler-Röhrborn A, Annas M, Faustmann P, Lennartz K, Zimmermann C W (1994). A standardized protocol for flow cytometric analysis of cells isolated from cerebrospinal fluid. *J. Neurol. Sci.* 121(1), 74-78.

6. Berhanu D, Mortari F, De Rosa SC and Roederer M (2003). Optimized lymphocyte isolation methods for analysis of chemokine receptor expression. *J. Immunol. Methods.* 279(1-2), 199-207.

7. Gratama J W, Kraan J, Van den Beemd R, Hooibrink B, Van Bockstaele D R, Hooijkaas H (1997). Analysis of variation in results of flow cytometric lymphocyte immunophenotyping in a multicenter study. *Cytometry.* 30(4), 166-177.

8. Gratama J W, Kraan J, Adriaansen H, Hooibrink B, Levering W, Reinders P, Van den Beemd M W, Van der Holt B, Bolhuis R L (1997). Reduction of interlaboratory variability in flow cytometric immunophenotyping by standardization of instrument set-up and calibration, and standard list mode data analysis. *Cytometry.* 30(1), 10-22.

9. Pattanapanyasat K, Kyle D E, Tongtawe P, Yongvanitchit K, Fucharoen S (1994). Flow cytometric immunophenotyping of lymphocyte subsets in samples that contain a high proportion of non-lymphoid cells. *Cytometry.* 18(4), 199-208.

10. Bedognetti D, Spivey T L, Zhao Y, Uccellini L, Tomei S, Dudley M E, Ascierto M L, De Giorgi V, Liu Q, Delogu L G, Sommariva M, Sertoli M R, Simon R, Wang E, Rosenberg S A, Marincola F M (2013). CXCR3/CCR5 pathways in metastatic melanoma patients treated with adoptive therapy and interleukin-2. *Br. J. Cancer.* 109(9), 2412-23.

11. Allaire M A, Tanné B, Côté S C and Dumais N (2013). Prostaglandin E 2 does not modulate CCR7 expression and functionality after differentiation of blood monocytes into macrophages. *Int J Inflam.* Doi: 10.1155/2013/918016.

12. McLinden R J, Labranche CC, Chenine A L, Polonis V R, Eller M A, Wieczorek L, Ochsenbauer C, Kappes J C, Perfetto S, Montefiori D C, Michael N L andKim J H (2013). Detection of HIV-1 neutralizing antibodies in a human CD4z/ CXCR4z/CCR5z T-lymphoblastoid cell assay system. *PLoS One.* 8(11), e77756.

13. Falk M K, Singh A, Faber C, Nissen M H, Hviid T, Sørensen T L (2014). Blood expression levels of chemokine receptor CCR3 and chemokine CCL11 in age-related macular degeneration: A case-control study. *BMC Ophthalmol.* Doi: 10.1186/1471-2415-14-22.

14. Noels H, Zhou B, Tilstam PV, Theelen W, Li X, Pawig L, Schmitz C, Akhtar S, Simsekyilmaz S, Shagdarsuren E, Schober A, Adams RH, Bernhagen J, Liehn EA, Döring Y, Weber C (2014). Deficiency of endothelial CXCR4 reduces reendothelialization and enhances neointimal hyperplasia after vascular injury in atherosclerosis-prone mice. *Arterioscler. Thromb. Vasc. Biol.* 34(6):1209-1220.

15. Ma N, Pang H, Shen W, Zhang F, Cui Z, Wang J, Wang J, Liu L, Zhang H (2015). Down regulation of CXCR4 by SDF-KDEL in SBC-5 cells inhibits their migration *in vitro* and organ metastasis *in vivo*. *Int. J. Mol. Med.* 35(2), 425-32.

Protocol No. 7.5: Detection of Nuclear Factor-Kappa β (NF-kB) Cellular Expression in Irradiated and Radioprotective Drug Treated Cells using Confocal Microscopy

Introduction

Nuclear factor kappa β (NF-kB) is a well known transcription factor that involved in the regulation of several cytokinesincluding interleukin (IL-1, IL-2, IL-6, IL-8), tumor necrosis factor alpha (TNF-α) and interferon-gamma (IFN-γ), adhesion molecules and immunoglobulin (Ig) genes. NFkB also play regulatory role in the inflammatory responses induction and cascading lead to apoptosis in cancer cells.The NFkB transcription factor is a complex of Rel A (a 65-kDa DNA binding subunit) and an associated 50-kDa protein (NFkB). In most cell types,except in B-cell lineage, the p50/p65 heterodimeric complex is localized in the cytoplasm, in a close association with its inhibitor, IkB.This inhibitory complex (NFkB-IkB) blocks nuclear translocation of NFkB. Phosphorylation and proteolysis of IkB allow free NF-kB p65 subunit to move towards nucleus. After binding to the specific promoterand associated transcriptional cofactor CBP, NFkB functions as a transcriptional activator (1-12).

Assay Requirement

RPMI-1640/specific cell culture medium, cell culture facilities, microscope slide, centrifuge, PFA, phosphate-buffered saline, methanol, Igepal, Nonidet P40, bovine serum albumin (BSA), NaN_3, goat anti-mouse NF-kB MAbs, FITC conjugated secondary antibodies, swine anti-goat antibody, Pipes-Triton, Oregon Green 488 Phalloidin, PI, Hepes buffer confocal microscope with 63 Plan Apochromat objective, gamma irradiator.

7.5.1 Assay Procedure

7.5.1.1 Detection of NF-kB by Immunofluorescence using Confocal Microscopy

7.5.1.1.1 Cell were grown in appropriate culture medium and divided in to following four experimental groups:

Gp A: Control cells (untreated)

Gp B: Radioprotective drug treated cells

Gp C: Cells irradiated by gamma radiation

Gp D: Irradiated cells pretreated with radioprotective drug

Followed by drug and radiation treatment, cells were allowed to grow for 6-12 h with 5 per cent CO_2 concentration and 95 per cent humidity. Cells were harvested and layered on a microscope slide. Following two procedures of cell fixation and permeabilization were used:

7.5.1.2 Cell Fixation using Paraformaldehyde (PFA)-methanol Treatment Procedure

Cells were xed with 4 per cent Paraformaldehyde (PFA) in phosphate-buffered saline (PBS) for 15 minutes at room temperature, washed with PBS, and transferred into 100 per cent methanol at -20°C for 5 minutes.

7.5.1.2.1 Methanol-igepal treatment procedure

Cells were incubated in 100 per cent methanol at -20°C for 5 minutes, washed with PBS and permeabilized by incubating them for 5 minutes at room temperature with PBS containing 2 per cent Igepal. Igepal is a non-ionic detergent that is chemically in distinguishable from Nonidet P40.

7.5.1.2.2 After fixation and permeabilization, the cells were washed twice with PBS containing1 per cent bovine serum albumin (BSA) and 0.01 per cent NaN_3(PBS-BSA).

7.5.1.2.3 Cells were then incubated for1 h at room temperature with 10-µg/ml anti-NF-kB monoclonal antibodies or polyclonal antibodies.

7.5.1.2.4 Followed by incubation, cells were washed twice in PBS-BSA and incubated with 150 µl of 1:20 FITC conjugated secondary antibodies in PBS-BSA.

7.5.1.2.5 Slides were analyzed by confocal microscopy and observations recorded.

7.5.1.2 Visualization of NFkB Localization in the Nuclei using Confocal Microscopy

7.5.1.2.1 To detect NF-kB translocation in cultured cells using flow cytometry, cells were cultured in 96-well plates.

7.5.1.2.2 Cells were cultured at a concentration of 2×10^5 cells per well, in the absence orpresence of varying concentrations of radioprotective drug (from 1 to 100µg/ml), and incubated for 30 minutes at 37°C.

7.5.1.2.3 Followed by incubation, cells were transferred to V bottom 96-well plates and washed twice with PBS.

7.5.1.2.4 Pure nuclei preparations were achieved by incubating the cells with 200 µl Pipes-Triton buffer (10 mM Pipes, 0.1 MNaCl, 2 mM $MgCl_2$ and 0.1 per cent Triton X100 for 30 minutes at 4°C.

7.5.1.2.5 Followed by two washes in PBS-BSA, nuclei were stained with anti–NF-kB antibodies (Monoclonal/polyclonal antibodies) at 5 µg/ml concentration for 30 minutes at 4°C.

7.5.1.2.6 After two more washes, nuclei, were incubated for 30 minutes at 4°C with therespective FITC-conjugated anti-Ig antibodies (all at 1/300dilution) and then washed twice in PBS-BSA.The prapations were then visuvalized under confocal microscopy.

7.5.1.2.7 To verify the purity of nuclei, preparations were stained with 50 nM Oregon Green 488 Phalloidin and PI (0.1 g/ml) and then visualized under confocal microscopy (1-12).

References

1. Blaecke A, Delneste Y, Herbault N, Jeannin P, Bonnefoy J Y, Beck A and J P Aubry (2002). Measurement of nuclear factor-kappa B translocation on lipopolysaccharide-activated human dendritic cells by confocal microscopy and flow cytometry. *Cytometry.* 48: 71–79

2. Bhattacharyya S, Borthakur A, Anbazhagan A N, Katyal S, Dudeja P K and Tobacman J K (2011). Specific effects of BCL10 serine mutations on phosphorylations in canonical and noncanonical pathways of NF-κB activation following carrageenan. *Am. J. Physiol. Gastrointest. Liver Physiol.* 301(3),475-486.

3. Noursadeghi M, Tsang J, Haustein T, Miller R F, Chain B M, Katz D R (2008). Quantitative imaging assay for NF-kappaB nuclear translocation in primary human macrophages. *J. Immunol. Methods.* 329(1-2), 194-200.

4. Molestina R E and Sinai A P (2005). Detection of a novel parasite kinase activity at the *Toxoplasma gondii* parasitophorous vacuole membrane capable of phosphorylating host Ikappa Balpha. *Cell Microbiol.* 7(3), 351-362.

5. Kiss-Toth E, Guesdon F M, Wyllie D H, Qwarnstrom E E and Dower S K (2000). A novel mammalian expression screen exploiting green fluorescent protein-based transcription detection in single cells. *J. Immunol. Methods.* 239(1-2),125-135.

6. Zhang Q, Wang J, Duan M T, Han S P, Zeng X Y, Wang J Y (2013). NF-κB, ERK, p38 MAPK and JNK contribute to the initiation and/or maintenance of mechanical allodynia induced by tumor necrosis factor-alpha in the red nucleus. *Brain Res. Bull.* 99, 132-39.

7. Wang P, Zhang Y, Xu K, Li Q, Zhang H, Guo J, Pang D, Cheng Y and Lei H (2013). A new ligustrazine derivative–pharmacokinetic evaluation and antitumor activity by suppression of NF-kappaB/p65 and COX-2 expression in S180 mice. *Pharmazie.* 68(9), 782-789.

8. Dong X Y, Liu W J, Zhao M Q, Wang J Y, Pei J J, Luo Y W, Ju C M and Chen J D (2013). Classical swine fever virus triggers RIG-I and MDA5-dependent signaling pathway to IRF-3 and NF-κB activation to promote secretion of interferon and inflammatory cytokines in porcine alveolar macrophages. *Virol. J.* Doi: 10.1186/1743-422X-10-286.

9. Bhattacharya N, Sarno A, Idler IS, Führer M, Zenz T, Döhner H, Stilgenbauer S and Mertens D (2010). High-throughput detection of nuclear factor-kappaβ

activity using a sensitive oligo-based chemiluminescent enzyme-linked immunosorbent assay. *Int. J. Cancer.* 127(2), 404-405.

10. Yan M, Xu Q, Zhang P, Zhou XJ, Zhang ZY and Chen WT (2010).Correlation of NF-kappaβ signal pathway with tumor metastasis of human head and neck squamous cell carcinoma. *BMC Cancer.* Doi: 10.1186/1471-2407-10-437.

11. Silva G E, Costa R S, Ravinal R C, Ramalho L Z, Dos Reis M A, Coimbra T M and Dantas M (2011). NF-kβ expression in IgA nephropathy outcome. *Dis Markers.* 31(1), 9-15.

12. Koshimizu J Y, Beltrame F L, de Pizzol J P Jr, Cerri P S, Caneguim B H, Sasso-Cerri E (2013). NF-kβ overexpression and decreased immunoexpression of AR in the muscular layer is related to structural damages and apoptosis in cimetidine-treated rat vas deferens. *Rep. Biol. Endocrinol.* Doi: 10.1186/1477-7827-11-29.

Protocol No. 7.6: Detection of Nuclear Factor-Kappa β (NF-kB) Cellular Expression in Irradiated and Radioprotective Drug Treated Cells using Flow Cytometery

Assay Requirement

RPMI-1640 cell culture medium, cell culture facilities, microscope slide, centrifuge, PFA, phosphate-buffered saline, methanol, Igepal, Nonidet P40, bovine serum albumin (BSA), NaN$_3$, goat anti-mouse NF-kB MAbs, FITC conjugated secondary antibodies, Pipes-Triton, Oregon Green 488 Phalloidin, PI, Hepes buffer, flow cytometer, gamma irradiator.

7.6.1 Flow Cytometry Analysis of NF-kB Translocation

7.6.1.2.1 To detect NF-kB translocation in cultured cells using flow cytometry, cells were cultured in as bottom 96-well plates.

7.6.1.2.2 Cells were cultured at a concentrationof 2×10^5 cells/well, in the absence or presence of varying concentrations of radioprotective drug (from 1 to 100 µg/ml) and incubated for 30 minutes at 37°C.

7.6.1.2.3 Followed by incubation, cells were transferred to V bottom 96-well plates and washed twice with PBS.

7.6.1.2.4 Pure nuclei preparations were achieved by incubating the cells with 200 µl Pipes-Triton buffer (10 mM Pipes, 0.1 M NaCl, 2 mM MgCl$_2$ and 0.1 per cent Triton X 100 for 30 minutes at 4°C.

7.5.1.2.5 Followed by two washes in PBS-BSA, nuclei were stained withanti-NF-kB antibodies (Monoclonal/polyclonal antibodies) at 5 µg/ml concentration for 30 minutes at 4°C.

7.6.1.2.6 After two more washes, nuclei were incubated for 30 minutes at 4°C with the respective FITC-conjugated anti-Ig antibodies (all at 1/300 dilution) and then washed twice in PBS-BSA. The preparation was then visuvalized under confocal microscopy.

7.6.1.2.7 Cells/Nuclei preparations were analyzed by flow cytometry equipped with the pulse processor module, permitting the doublet discrimination.

7.6.1.2.8 Nuclei were counter stained with 1µg/ml propidium iodide. Single nuclei were gated on the basis of PI staining (FL-3 measured at 630 nm), after doublet elimination by FL-3 area versus FL-3 width measurement.

7.6.1.2.9 Voltage settings on FITC parameters were done on isotype control samples. 2–3 x 10^4 events were recorded for each sample.

References

1. Blaecke A, Delneste Y, Herbault N, Jeannin P, Bonnefoy J Y, Beck A and Aubry J P (2002). Measurement of nuclear factor-Kappa β translocation on lipopolysaccharide-activated human dendritic cells by confocal microscopy and flow cytometry. *Cytometry*. 48, 71–79.

2. Noursadeghi M, Tsang J, Haustein T, Miller R F, Chain B M andKatz D R (2008). Quantitative imaging assay for NF-kappaβ nuclear translocation in primary human macrophages. *J. Immunol. Methods*. 329(1-2),194-200.

3. Zhang Q, Wang J, Duan M T, Han S P, Zeng X Y and Wang J Y (2013). NF-κB, ERK, p38 MAPK and JNK contribute to the initiation and/or maintenance of mechanical allodynia induced by tumor necrosis factor-alpha in the red nucleus. *Brain Res. Bull*. 99,132-39.

4. Ling X Q and Wang J K (2013). Techniques for assaying the activity of transcription factor NF-κB. *Yi Chuan*. 35(5):551-70.

5. Fluhr H, Spratte J, Bredow M, Heidrich S and Zygmunt M (2013). Constitutive activity of Erk1/2 and NF-κB protects human endometrial stromal cells from death receptor-mediated apoptosis. *Reprod Biol*. 13(2),113-121.

6. Yan M, Xu Q, Zhang P, Zhou XJ, Zhang ZY and Chen WT (2010). Correlation of NF-kappa β signal pathway with tumor metastasis of human head and neck squamous cell carcinoma. *BMC Cancer*. Doi: 10.1186/1471-2407-10-437.

Protocol No. 7.7: Flow Cytometric Detection of Intracellular Th1/Th2 Cytokines using Whole Blood Collected from the Irradiated and Radioprotective Drug Treated Mice

Introduction

Cytokines are the prominent protein biomarkers of modulated immune response under various pathological conditions. Type 1 cytokines particularly IFN-γ, interleukin (IL)-12, and tumor necrosis factor-α specifically promote pro-inflammatory immune responses, whereas, type 2 cytokines such as IL-4, IL-5, IL-10, and IL-13) induce anti-inflammatory, antibody dependent immune responses. Purturbed type 1/type 2 lineage cytokine production lead to twisted memory T-helper 1 (Th1) or T-helper 2 (Th2) development, which secrete type 1/type 2 cytokines, respectively.Various cytokines assays like single and multiplexed ELISAs reverse transcription-PCR, Taq man real-time PCR and immunohistochemistry etc. are available. However, ELISA assay is most commonly employed in most of the cases due to its simplicity. However, the instant uptake of released cytokines by surrounding immune cells requires immediate collection and processing at appropriate times. One more limitation of ELISAs assay is that it cannot unreavel the cellular source of the cytokines secreted into blood plasma or serum. Prsesent flow cytometry based method circumvents these limitations and showed great potential for biomonitoring of immune cells functions in human populations. Flow cytometry based detection of intracellular cytokines is a functional assay that measures the ability of specific immune cells to express type 1 and type 2 cytokines after polyclonal stimulation with mitogens.

Assay Requirement

Cell culture facility, blood sample collection tubes, fluorescence-activated cell sorting lysing solution, centrifuge, permeabilizing solution, bovine serum albumin, NaN_3 PBS, IFNγ-FITC antibodies, IL-4-PE antibodies, CD4, specific antibodies, washing buffer, paraformaldehyde, CD4+ antibodies, flow cytometer, gamma irradiator.

7.7.1 Assay Procedure

7.7.1.1 Animals or cells were divided into following four groups:

Gp.1 Control animals or cells (untreated animals or cells)

Gp.2 Irradiated animals or cells (animals or cells treated with gamma radiation)

Gp.3 Animals or cells treated with radioprotective drug

Gp.4 Irradiated animal or cells pretreated with radioprotective drug

Venous blood samples were collected from the treated animals at different time intervals and stored at room temperature in the dark till use for further experimentation. However, if cultured cells are under investigation, cell free suspension of them can be used for flow cytometric analysis.

7.7.1.2 **Blood Cell Culture**

7.7.1.2.1 Blood samples were collected from experimental animals and cultured at 2, 24, 48 and 72h.

7.7.1.2.2 The cultured cells were lysed with 1x fluorescence-activated cell sorting lysing solution (3 ml) for 8 minutes.

7.7.1.2.3 The cells were then centrifuged at 2000 rpm for 5 minutes at 4°C.

7.7.1.2.4 The supernatant aspirated and 1x permeabilizing solution (500µl) was mixed into the pellet.

7.7.1.2.5 Mixture was incubated for 10 minutes at room temperature in the dark.

7.7.1.2.6 Followed by washing with washing buffer (3ml) (1 per cent bovine serum albumin, 0.1 per cent NaN_3, 1x PBS), cytokine-specific antibodies (20-30µl; *i.e.* IFNγ-FITC, IL-4-PE) were added to the cells and cells suspension incubated for 30 minutes at room temperature.

7.7.1.2.7 Cells were washed with washing buffer (composition mentioned above) and again resuspended in 1 per cent paraformaldehyde (500µl) and stored at 4 °C until flow cytometry analysis.

7.7.1.2.8 Percentage of Th1 and Th2 cytokine-producing cells in the total population of CD4+ T-helper cells were calculated in terms of the number of IFNγ positive and IL-4 possitive cells respectuively.

7.7.1.2.9 A minimum of 5000 -6000 CD4+ cells need to be counted from each sample (1-7).

References

1. Duramad P, Nina T. Christopher H, McMahon W, Hubbard A and Eskenazi B (2004). Flow cytometric detection of intracellular Th1/Th2 cytokines using whole blood: Validation of immunologic biomarker for use in epidemiologic studies. *Cancer Epidemiol. Biomarkers Prev.* 13 (9), 1452-1458.

2. Pinto R A, Arredondo S M, Bono M R, Gaggero A Aand Díaz P V (2006).T helper 1/T helper 2 cytokine imbalance in respiratory syncytial virus infection is associated with increased endogenous plasma cortisol. *Pediatrics.* 117(5), e878-86.

3. Hassan M A, Eldin A M and Ahmed M M (2008). T-helper2/T-helper1 imbalance in respiratory syncytial virus bronchiolitis in relation to disease severity and outcome. *Egypt J. Immunol.* 15(2): 153-160.

4. Kasakura S (1998). A role for T-helper type 1 and type 2 cytokines in the pathogenesis of various human diseases. *Rinsho Byori.* 46(9), 915-921.

5. Skapenko A and Schulze-Koops H (2007). Analysis of Th1/Th2 T-cell subsets. *Methods Mol. Med.* 136, 87-96.

6. Soltész P, Aleksza M, Antal-Szalmás P, Lakos G, Szegedi G and Kiss E (2002). Plasmapheresis modulates Th1/Th2 imbalance in patients with systemic lupus erythematosus according to measurement of intracytoplasmic cytokines. *Autoimmunity.* 35(1), 51-56.

7. Kuo M L, Huang J L, Yeh K W, Li P S andHsieh K H (2001). Evaluation of Th1/Th2 ratio and cytokine production profile during acute exacerbation and convalescence in asthmatic children. *Ann. Allergy Asthma Immunol.* 86(3), 272-276.

Protocol No. 7.8: Determination of Micronuclei Frequency in Bone Marrow Cells of Irradiated and Radioprotective Drug Treated Mice

Assay Requirement

Bovine serum albumin, Giemsa dye, microcenterifuge tube (1.5 ml), syrenges (1ml), referegerated centrifuge, gamma irradiator.

7.8.1 Assay Procedure

7.8.1.1 Animals were divided into four experimental groups:

    i. Control untreated mice

    ii. Mice treated with Radioprotective drug

    iii. Irradiated mice

    iv. Irradiated mice pretreated by radioprotective drug

    Followed by various radiation and drug treatment, the mice were sacrificed and femur bone was dissected out.

7.8.1.2 The bone marrow cells flushed out from femur were collected in a suspension in 1.5 ml centrifuge tube containing 5 per cent BSA.

7.8.1.3 Single cell suspension of the bone marrow cells was prepared by gentle shaking and pipetting. Cell pellet was collected by centrifugation at 2000 rpm for 5 minutes at 4°C.

7.8.1.4 The bone marrow cells pellet was resuspended in a drop of 5 per cent BSA and cells smear were prepared. After air drying the smear slides were stained with Giemsa dye.

7.8.1.5 Micronucleated polychromatic erythrocytes and Non-chromatic erythrocytes were observed under light Microscope.

7.8.1.6 To estimate the effect of gamma radiation and its modulation by radioprotective drug treatment, percentage (per cent) of micronucleated polychromatic erythrocytes (MnPCEs), micronucleated non-chromatic erythrocytes (MnNCEs) and the ratio of PCE Vs (PCE + NCE) cells was calculated (1-7).

References

1. Barangi S, Vidya S M, GSanjeev, Vaman R C, Rajesh KP and Vinutha K (2012). Evaluation of radioprotective efficacy of *Ficus racemosa* in Swiss albino mice exposed to electron beam radiation. *J. Biochem. Tech.* 3(5), S212-S217.

2. Chang Y, Zhou C, Huang F, Torous DK, Luan Y, Shi C, Wang H, Wang X, Wei N, Xia Z, Zhong Z, Zhang M, An F, Cao Y, Geng X, Jiang Y, Ju Q, Yu Y, Zhu J, Dertinger SD, Li B, Liao M, Yuan B, Zhang T, Yu J, Zhang Z, Wang Q and Ma J (2014). Inter-laboratory validation of the *in vivo* flow cytometric micronucleus analysis method (MicroFlow®) in China. *Mutat. Res. Genet. Toxicol. Environ. Mutagen.* 15, 772:6-13

3. Belmont-Díaz J, López-Gordillo A P, Molina Garduño E, Serrano-García L, Coballase-Urrutia E, Cárdenas-Rodríguez N, Arellano-Aguilar O and -Montero-Montoya R D (2014). Micronuclei in bone marrow and liver in relation to hepatic metabolism and antioxidant response due to coexposure to chloroform, dichloromethane and toluene in the rat model. *Biomed. Res. Int.* doi: 10.1155/2014/425070.

4. Zhu N, Li H, Li G and Sang N (2013). Coking wastewater increases micronucleus frequency in mouse *in vivo* via oxidative stress. *J Environ Sci* (China). 2013 25(10), 2123-2139.

5. Kawasaki I, Suzuki Y and Yanagisawa H (2013). Zinc deficiency enhances the induction of micronuclei and 8-hydroxy-2'-deoxyguanosine via superoxide radical in bone marrow of zinc-deficient rats. *Biol. Trace Elem. Res.* 154(1), 120-126.

6. Alghazal M A, Sutiaková I, Kovalkovicová N, Legáth J, Falis M, Pistl J, Sabo R, Benová K, Sabová L and Váczi P (2008). Induction of micronuclei in rat bone marrow after chronic exposure to lead acetate trihydrate.Toxicol Ind Health. 24(9), 587-593.

7. Ran Y, Wang R, Lin F, Hasan M, Jia Q, Tang B, Xia Y, Shan S, Wang X, Li Q, Deng Y and Qing H (2014). Radioprotective effects of Dragon's blood and its extract against gamma irradiation in mouse bone marrow cells. *Phys. Med.* 30(4), 427-431.

Protocol No. 7.9: Determination of Hematopoietic Systems Status by Estimation of T Cells, B Cells and Monocyte/Macrophages Cells Population using Cell Surface Marker Expression in Irradiated Mice and its Modulation in Radioprotective Drug Pretreated Mice using Flow Cytometry

Assay Requirement

Blood collection tube with anticoagulant, PBS, PAF, CD19-FITC, CD14-FITC, CD3-FITC, saponin, 7AAD, flow cytometer, gamma irradiator.

7.9.1 Assay Procedure

7.9.1.1 Animals were divided into four experimental groups:

    i. Control untreated mice

    ii. Mice treated with radioprotective drug

    iii. Irradiated mice

    iv. Irradiated mice pretreated by radioprotective drug

Followed by various radiation and drug treatment, blood was collected from the sinus orbital of the mice and PBMC separated using ficol gradient centrifugation method.

7.9.1.2 PBMCs were washed with PBS and fixed with PBS/PAF1 per cent at room temperature for 20 minutes.

7.9.1.3 PBMCs sub-populations were labelled by following FITC lebelled monoclonal antibodies (5µg/10 cells) to analysed T cells, B cells and macrophages in the blood of the irradiated and drug treated mice.

    i. CD19-FITC for B-cells,

    ii. CD14-FITC for monocytes/macrophages

    iii. CD3-FITC for T-lymphocytes, in separated tubes at 4°C for 45 minutes.

7.9.1.4 Followed by incubation, cells were washed twice with PBS and permeabilized by 0.03 per cent saponin at room temperature for 15 minutes.

7.9.1.5 Simultaneously, PBMCs DNA can also be counterstained with 7AAD (0.25µg/10cells) at room temperature for 10 minutes.

7.9.1.6 At the last after final washing in PBS-FBS, PBMCs sub-populations were analyzed by flow cytometry.

7.9.1.7 A total of 10^4 to 10^6 events should be recorded for each sample (1-7).

References

1. Lafarg S, Hamzeh-Cognasse H, Chavarin P, Genin C, Garraudand O, Cognasse F (2007). A flow cytometry technique to study intracellular signals NF-kB and STAT3 in peripheral blood mononuclear cells. *BMC Mol. Biol.* Doi:10.1186/1471-2199-8-64.

2. Kamada H, Taki S, Nagano K, Inoue M, Ando D, Mukai Y, Higashisaka K, Yoshioka Y, Tsutsumi Y and Tsunoda S I (2014). Generation and characterization of a bispecific diabody targeting both EPH receptor A10 and CD3. *Biochem Biophys Res Commun.* Doi: 10.1016/j.bbrc.2014.12.030.

3. Elizei S S, Poormasjedi-Meibod M S, Li Y, Jalili R B and Ghahary A (2014). Effects of Kynurenine on CD^{3+} and macrophages in wound healing. *Med Arch.* 68(4), 236-238.

4. Karamehic J, Zecevic L, Resic H, Jukic M, Jukic T, Ridjic O and Panjeta M (2014). Immunophenotype lymphocyte of peripheral blood in patients with psoriasis. *Coric J.* 68(4), 236-238.

5. Longshan L, Dongwei L, Qian F, Jun L, Suxiong D, Yitao Z, Yunyi X, Huiting H, Lizhong C, Jiguang F and Changxi W (2014). Dynamic analysis of B-cell subsets in de novo living related kidney transplantation with induction therapy of basiliximab. *Transplant Proc.* 46(2), 363-367.

6. Rotti H, Guruprasad KP, Nayak J, Kabekkodu SP, Kukreja H, Mallya S, Nayak J, Bhradwaj R C, Gangadharan G G, Prasanna B V, Raval R, Kamath A, Gopinath PM, Kondaiah P, Satyamoorthy K (2014). Immunophenotyping of normal individuals classified on the basis of human dosha prakriti. *J. Ayurveda Integr. Med.* 5(1), 43-49.

7. Jabbar K J, Medeiros L J, Wang S A, Miranda R N, Johnson M R, Verstovsek S, Jorgensen J L (2014). Flow cytometric immunophenotypic analysis of systemic mastocytosis involving bone marrow. *Arch. Pathol. Lab. Med.* 138(9), 210-1214.

Index